SAFETY AND RISK IN SOCIETY SERIE…

ACCIDENTS: CAUSES, ANALYSIS AND PREVENTION

SAFETY AND RISK IN SOCIETY SERIES

Chemical Safety Board
Robert W. Talford (Editor)
2009. ISBN 978-1-60692-586-7

Consumer Product Safety Issues
Peter A. Varga and Mate D. Pinter
2009. ISMN 978-1-60456-826-4

Fire in the United States: A Reference Work on the Nation's Fire Problem
Ciro Bouchard and Marco Fortin (Editors)
2009. ISBN: 978-1-60741-746-0

Protective Devices: Types, Uses and Safety
Victor E. Argosyan (Editor)
2009. ISBN: 978-1-60876-223-1

Risk Management
Benigno Jordão and Emilio Sousa (Editors)
2009. ISBN: 978-1-60876-011-4

Fire Safety
Ingmar Søgaard and Hans Krogh (Editors)
2009. ISBN: 978-1-60741-490-2

Accidents: Causes, Analysis and Prevention
Henri Bédard and Geraud Delashmit (Editors)
2009. ISBN 978-1-60741-712-5

SAFETY AND RISK IN SOCIETY SERIES

ACCIDENTS: CAUSES, ANALYSIS AND PREVENTION

HENRI BÉDARD
AND
GERAUD DELASHMIT
EDITORS

Nova Science Publishers, Inc.
New York

LIBRARY OF CONGRESS CATALOGING-IN-PUBLICATION DATA

Accidents : causes, analysis, and prevention / [compiled by] Henri Bédard and Geraud Delashmit.
p. cm.
Includes index.
ISBN 978-1-60741-712-5 (hardcover)
1. Traffic safety. 2. Traffic accident investigation. 3. Automobile driving--Physiological aspects. I. Bédard, Henri. II. Delashmit, Geraud.
HE5614.A324 2009
363.12'5--dc22

2009029338

Publishers, Inc. New York

CONTENTS

PREFACE

In the last decade, nearly every state in the U.S. has introduced a Graduated Driver Licensing (GDL) program to limit the exposure of teenage drivers to high-risk conditions during the first years after licensure. This book evaluates the effectiveness of GDL programs in reducing teen traffic fatalities over the period between 1992 and 2006. This book also investigates the psychological functions of driving and risky driving as perceived by adolescents and the factors underlying psychological functions of driving and risky driving. Furthermore, the impact of accident duration on traffic congestion is also analyzed, as well as the multiple factors that may affect driving performance. Other chapters in this book examine the injury risks experienced by vulnerable road users and challenges in obtaining accurate information on the magnitude and circumstances of their crashes, studies where human factors related to occupational accidents are analyzed, and an evaluation of the factors that are associated with the risk perception to have night-time car crashes in young drivers, as well as the strategies most commonly used to counteract sleepiness at the wheels. Implications of the study on road safety and recommendations for future research are also discussed.

Chapter 1 - Teenage novice drivers are involved in traffic accidents more than any other age group, in particular during their first year of unsupervised driving. Increasing their driving experience on the one hand reduces their risk of crash involvement, but on the other hand raises their exposure to risk. In the last decade, nearly every state in the U.S. has introduced a Graduated Driver Licensing (GDL) program to limit the exposure of teenage drivers to high-risk conditions during the first years after licensure.

This chapter evaluates the effectiveness of GDL programs in reducing teen traffic fatalities over the period between 1992 and 2006. Count data models are estimated for panel data, where traffic-related fatalities are dependent variables calculated per state and per year for various age cohorts. Model specifications include indicators that specify the implementation of GDL programs within a state during a certain year, as well as exogenous variables that control for other potentially relevant determinants of fatality reduction, exposure to risk in different states, and macroeconomic factors that possibly influence crash rates.

Results show that GDL programs have been quite effective in reducing traffic fatalities among 15-17-year-olds. Specifically, the average graduated licensing system is evaluated to decrease teen fatality counts by 11.1%. In addition, results illustrate that the most restrictive policies – according to the taxonomy of the Insurance Institute for Highway Safety – have been the most effective. “Good” programs are estimated to reduce teen traffic fatalities by

15.1%, "fair" programs by 7.8%, and "marginal" programs by 5.1%. Last, results suggest that GDL regulations do not have traffic-safety implications for young drivers who have already reached full licensure, and that even the effect on 17-year-olds is inferior to the effect on 15- and 16-year-olds. In conclusion, the conceptual goal of protecting young novice drivers while they are gaining experience appears to have been achieved to a significant extent, in particular during the first, critical year after licensure.

Chapter 2 - Driving a motor vehicle is a key behavior in the development of adolescents; however, driving may be dangerous for teenagers, especially when they fail to respect the traffic code or drive in a risky manner. Risky driving is, in fact, one of the main causes of motor vehicle accidents. Psychological literature has stressed that driving and risky driving in adolescence represent self-regulated actions, oriented towards the attainment of meaningful goals for individual development. In other words, driving and risky driving fulfill psychological functions in relation to adolescent developmental tasks (Jessor, 1998; Bonino et al., 2005). The study of these psychological functions is crucial to understand why adolescents engage in risky driving.

The aims of the study were to investigate: 1) the psychological functions of driving and risky driving as perceived by adolescents; 2) the factors underlying psychological functions of driving and risky driving; 3) the relationships between driving and risky driving functions and involvement in risky driving. With reference to previous studies (Bonino, et al. 2005), two factors underlying the functions of risky driving were expected, the first includes functions related to identity definition (experimentation and perception of control, transgression, sensation seeking, coping) and the second includes functions related to peer relationships (visibility, peer emulation, surpassing peers, identification with peers). Moreover, different levels of involvement in risky driving were expected in relation to different perceived functions.

Data were collected through an anonymous self-report questionnaire. The sample included 626 adolescents ages 15 to 19; it was balanced for gender and representative of adolescents attending high schools in Turin (northwestern Italy). Risky driving was evaluated using the scale by Bonino et al. (2005), while scales of psychological functions were defined on the basis of a preliminary explorative study using focus group interviews with adolescent drivers (Bina et al., 2008). Confirmatory factor analysis identified the factors underlying the psychological functions of driving and risky driving. Hierarchical multiple regression analysis investigated the relationships between functions and involvement in risky driving.

Factor analysis revealed two factors underlying driving functions: leisure (driving for fun, to meet friends, and to pass the time) and autonomy (in travel, from parents, to reach distant places) and two factors underlying risky driving functions (identity definition and peer relationships). Regression analysis showed that only identity definition functions were linked to involvement in risky driving. Driving functions (leisure and autonomy) and relationship functions of risky driving were not statistically significant in the regression model.

Implications for risky driving and motor vehicle accident prevention programs are discussed.

Chapter 3 - Developmental psychology literature has stressed the role of parents' and friends' behavior models and attitudes in adolescents' involvement in risk behavior (Jessor et al., 1991; Bonino et al., 2005). With reference to risky driving (traffic code violations and driving in a risky way) different studies have investigated the role of parents' and friends' driving style as well as their attitudes about adolescent involvement in risky driving (Hartos et al., 2000; Taubman-Ben-Ari et al., 2005; Wilson et al., 2006). However, previous studies have only examined these predictors separately. The present study investigates the combination of these variables in the prediction of self-reported risky driving, by taking into account the differences between the maternal and paternal role and the roles of friends and the best friend. Evaluations of parents' and friends' risky driving models and their disapproval of risky driving are based on adolescents' perceptions.

The main aim of the study was to investigate the relationship between parents' risky driving models and risky driving disapproval (looking separately at mothers and fathers), friends' risky driving model and risky driving disapproval (distinguishing between friends and best friend) and adolescents' involvement in risky driving. Gender differences were also investigated. We expected that adolescents' greater involvement in risky driving was related to parents' and friends' risky driving model, while lower involvement was related to parents' and friends' disapproval of risky driving.

Data were collected through an anonymous self-report questionnaire. The study involved 341 adolescent drivers (car, motorbike, moped) (50% boys), ages 15 to 19, attending different high schools in Piedmont (northwestern Italy). Risky driving was evaluated using the scale developed by Bonino et al. (2005), while items related to parents' and friends' risky driving model and risky driving disapproval were formulated specifically for this study. Correlations and hierarchical multiple regressions were carried out, respondent age was controlled for, and gender and average weekly miles driven were considered in order to investigate relationships between variables.

Main results stressed that: 1) in the total sample significant predictors of risky driving were gender, paternal, friends' and best friend's risky driving models, and friends' disapproval of risky driving; 2) for boys the best predictor was friends' disapproval of risky driving followed by friends' and paternal risky driving model; 3) for girls significant predictors were age and best friend's risky driving model.

Implications for risky driving prevention in adolescence are discussed.

Chapter 4 - Vulnerable road users, namely, motorcyclists, bicyclists and pedestrians, contribute significantly to morbidity and mortality rates for road casualties. Many of these injuries occur off-road without police attendance or do not involve a collision with a motor vehicle. As a result, the magnitude and scope of their injuries are often underestimated. Behavioural factors also play a pivotal role, but information on the prevalence of behaviours related to an increased risk of involvement in a crash is also rather limited.

A six-month prospective study of vulnerable road users presenting to emergency departments due to an injury from a crash or non-motor vehicle collision was conducted in Perth, Western Australia. Through a structured questionnaire, the characteristics of the road user, the injury sustained and the crash were described and the crash-related behaviour investigated.

Of the total 392 vulnerable road users, 282 (72%) participated in the interview with the majority being bicyclists (53%) followed by motorcyclists (37%). Males represented 80% of the presentations and dominated each road user type except pedestrians. Many bicyclists and motorcyclists were injured by a collision that did not involve a motor vehicle and/or occurred in off-road situations and was not attended by police. In terms of behavioural factors, 30% of pedestrians reported the motor vehicle involved was moving faster than initially thought, particularly among older adults aged 55 and above. The study found that 7% of motorcyclists and 23% of bicyclists were not wearing their helmet at all times, while one third of the adolescent bicyclists did not wear a helmet at the time of injury. Only 20% of bicyclists and 67% of motorcyclists injured during the evening were wearing visible clothing. Overall, 14% of vulnerable road users admitted being distracted and 13% of them were hurrying before the injury event. Excessive speed was a contributing factor in crashes sustained by motorcyclists.

In conclusion, data linkage between emergency presentations, hospital admissions and police records is recommended to improve the accuracy and representativeness of injury data. Behavioural factors play a pivotal role in the protection of vulnerable road users. Their safety, to a large extent, is determined by their risk taking behaviour. Safety promotion programs targeting the wearing of helmet and visible clothing should be aimed at both motorcyclists and bicyclists, especially among adolescent bicyclists.

Chapter 5 - It is generally accepted that 80-90% of accidents are due to human error. Two out of three fatal occupational accidents in Australia are caused by human factor. For accident prevention, a superficial explanation about a human factor as the cause of accident is enough; we need a more detailed analysis of the nature of human factors.

The aim of this review of literature is to examine studies, where human factors related to occupational accidents were analyzed. On the one hand, well-known theories of accident causes (e.g. accident proneness) were presented with new empirical results. On the other hand, new areas of research of human factors are presented like the relationship between safety climate and occupational accidents.

Risk taking is one important area of human factors leading to occupational accidents. Risk taking theories and empirical research in the area presented. The other significant area is attributions, especially self-defensive attributions. Human errors and safety management are also meaningful topics in this review.

Chapter 6 - Sleepiness is often a contributing factor in car crashes. Despite drivers under 30 years old are particularly prone to sleep-related accidents in the early morning, they often underestimate the role of sleepiness as a risk factor for car crashes. The aim of the study was to examine which factors are associated with the risk perception to have night-time car crashes in young drivers and to evaluate the strategies most commonly used to counteract sleepiness at the wheels. 1,123 young drivers (18-25 years old) with at least a 6-months driving experience participated to the study. Results indicated that males are less worried about night-time car crashes than females. The perceived likelihood of having a sleep-related car crash is low for those who report driving at night frequently. On the other hand, the level of concern regarding sleep-related car crashes increases in young people who had experienced sleepiness-impaired driving.

Chapter 7 - Highway congestion is seriously affected by traffic accidents. The stochastic nature associated with an accident causes significant difficulties in accident duration forecast, travel time prediction and route guidance. Therefore it is vital to gain an insight into the characteristics of accidents. In this study we analyze the accident database from the Police Radio Station in order to find useful clues with any possible data feature. Focus is mainly on the impact of accident duration on traffic congestion. Further studies could refer this study results to process the model inputs for model development.

Chapter 8 - The objective of the present research was to undertake a comparative and combined assessment of the three commonly used predictor categories of driving outcomes, namely demographic, information processing, and personality variables. Research participants (Study 1: n = 153; Study 2: n = 335) completed a predictor battery and a self-report measure of motor vehicle crashes and moving violations. Using a series of hierarchical regression analyses, the obtained results generally suggest that, consistent with the conceptual need to recognize the role of the "total" individual in driving-related behaviors, the use of multiple predictors from multiple predictor categories in the prediction of driving outcomes is empirically meaningful. As would be expected, this approach accounted for substantially more variance in the criteria than either any single predictor or individual predictor category would by itself. In addition, the findings particularly highlighted the importance of including multiple personality variables in models of driving outcomes. The implications of these findings and suggestions for future research are discussed. Information Processing, Personality, and Demographic Variables as Predictors of Crashes and Moving Violations.

Chapter 9 - This chapter reviews the literature on distracted driving and makes the case that the focus of the research and legislation designed to prevent accidents is often misdirected. For instance, as of now (or in the near future) the Insurance Institute for Highway Safety (2007) noted that only six U.S. states have bans on handheld cellular phone use among drivers, but none of these states ban the use of hands-free cellular phones. These laws appear to address accident prevention by ensuring that drivers have both hands free to operate their vehicles. However, recent studies have shown that visual and cognitive distractions (rather than occupied hands) are the main contributing factors to traffic accidents in which cellular phones were currently in use. In fact, several studies have shown that impaired driving performance is just as likely to occur whether the phone is held or hands-free (e.g., Horrey & Wickens, 2006). The current chapter argues that cellular phones (and other nonessential devices) contribute to traffic accidents by impairing drivers' situation awareness (SA). That is, conversing with an individual who is not in the car, and who is therefore unable to alter the conversation based on current driving conditions, impairs drivers' ability to identify, attend to, and quickly react to relevant environmental stimuli and situations. Though situation awareness has been studied extensively in military pilots, this research is only recently being applied to the population of drivers. The relevance of SA to safe driving and potential methods for maintaining SA are addressed. These methods include technological advances such as tactile cueing and warning systems, as well as often overlooked psychological approaches for improving concentration.

In: Accidents: Causes, Analysis and Prevention
Editors: Henri Bédard and Geraud Delashmit

ISBN: 978-1-60741-712-5

Chapter 1

REDUCING TEEN FATALITIES: DO GRADUATED LICENSING PROGRAMS HELP SOLVE THE PROBLEM?

Carlo Giacomo Prato
Ran Naor Road Safety Research Center, Technion – Israel Institute of Technology
Technion Campus, Haifa 32000, Israel

ABSTRACT

Teenage novice drivers are involved in traffic accidents more than any other age group, in particular during their first year of unsupervised driving. Increasing their driving experience on the one hand reduces their risk of crash involvement, but on the other hand raises their exposure to risk. In the last decade, nearly every state in the U.S. has introduced a Graduated Driver Licensing (GDL) program to limit the exposure of teenage drivers to high-risk conditions during the first years after licensure.

This chapter evaluates the effectiveness of GDL programs in reducing teen traffic fatalities over the period between 1992 and 2006. Count data models are estimated for panel data, where traffic-related fatalities are dependent variables calculated per state and per year for various age cohorts. Model specifications include indicators that specify the implementation of GDL programs within a state during a certain year, as well as exogenous variables that control for other potentially relevant determinants of fatality reduction, exposure to risk in different states, and macroeconomic factors that possibly influence crash rates.

Results show that GDL programs have been quite effective in reducing traffic fatalities among 15-17-year-olds. Specifically, the average graduated licensing system is evaluated to decrease teen fatality counts by 11.1%. In addition, results illustrate that the most restrictive policies – according to the taxonomy of the Insurance Institute for Highway Safety – have been the most effective. "Good" programs are estimated to reduce teen traffic fatalities by 15.1%, "fair" programs by 7.8%, and "marginal" programs by 5.1%. Last, results suggest that GDL regulations do not have traffic-safety implications for young drivers who have already reached full licensure, and that even the effect on 17-year-olds is inferior to the effect on 15- and 16-year-olds. In conclusion, the conceptual goal of protecting young novice drivers while they are gaining experience appears to have been achieved to a significant extent, in particular during the first, critical year after licensure.

1. INTRODUCTION

Teenage novice drivers are involved in traffic accidents more than any other age group, in particular during their first year of unsupervised driving (Preusser & Leaf, 2003; Williams, 2003). Relevant factors contributing to their high crash risk are immaturity and driving inexperience (e.g., Mayhew, 2003; Simons-Morton & Hartos, 2003), late night driving (e.g., Lin & Fearn, 2003; Simons-Morton & Hartos, 2003; Williams, 2007) and social pressure from the presence of teen passengers (e.g., Preusser et al., 1998; Chen et al., 2000; Lin & Fearn, 2003; Williams, 2007). Moreover, teenagers are less able to assess driving hazards than older drivers, tend to exhibit risk-prone behavior behind the wheel, and think that their crash risk is low (Ferguson, 2003). Clearly, teenage immaturity, driving inexperience and risk-taking behavior constitute a deadly mixture that significantly contributes to the over-representation of young drivers in road injuries and fatalities.

For young drivers, the paradox lies in the fact that they need to increase their driving experience to reduce the risk of crash involvement, but the more they drive, the more they expose themselves to risks. For policy makers, the dilemma lies in the fact that they need to allow teenage drivers to gain experience without overly exposing them to high crash risk. To this extent, every U.S. jurisdiction has introduced Graduated Driver Licensing (GDL) programs to limit the exposure of teenage drivers to high-risk conditions during the first years after licensure. GDL programs are three-stage systems consisting of a learner's permit, a provisional license, and a full license. Essential aspects of these systems are that a learner's permit allows driving only under supervision of a fully licensed person, a provisional license allows unsupervised driving under restrictive conditions, and both the learner's permit and the provisional license must be held for a predetermined period. The learner's permit addresses driving inexperience, hazard perception and immature behavior by providing learning opportunities in a large variety of situations. The provisional license concentrates on late night driving, teen passengers' presence, cell phone use, and seatbelt wearing by imposing restrictions during the initial months of unsupervised driving.

This chapter evaluates the effectiveness of GDL programs in reducing teen traffic fatalities over the period between 1992 and 2006. Information on licensing systems is obtained from the Insurance Institute for Highway Safety (IIHS, 2008), which maintains an updated collection of licensing laws in the United States. Information on fatal crashes is retrieved from the Fatality Analysis Reporting System (FARS) databases of the National Highway Traffic Safety Administration (NHTSA, 2008), which maintains a census of fatalities on U.S. public roads. Count data models are estimated for panel data, where traffic-related fatalities are dependent variables calculated per state and per year for various age cohorts. The econometric analyses presented in this chapter allow the examination of various questions related to the usefulness of GDL implementation.

The first question concerns the effectiveness of graduated licensing systems in reducing traffic-related fatalities among 15-17-year-olds. The specification of negative binomial models accounts for indicators of GDL implementation, as well as exogenous variables that control for other potentially relevant determinants of fatality reduction, exposure to risk in different states, and macroeconomic factors that possibly influence crash rates. With respect to previous multi-jurisdictional analyses of GDL effectiveness (e.g., Dee et al., 2005; Chen et al., 2006; Morrisey et al., 2006), the presented models analyze the most recent fatality data, expand the temporal scope of the research, and consider the most current legislation in order to provide a more exhaustive and up-to-date analysis.

The second question concerns the assumption of the greater effectiveness of GDL policies that impose the most restrictive rules in terms of length of the learner's period, duration of the intermediate phase and stringency of the restrictions. While maintaining the same control variables, model specification accounts for different indicators for GDL programs belonging to different classes according to the "good, fair, marginal and poor" taxonomy introduced by the IIHS. With respect to previous nationwide studies of GDL effectiveness (e.g., Baker et al., 2007; Dee et al., 2005; Chen et al., 2006; Morrisey et al., 2006), the presented models consider the most recent promulgations and the most current modifications of GDL laws, and in particular benefit from a larger number of state-by-year observations with "good" programs implemented.

The third question concerns the hypothesis of extension of life-saving benefits to young adults who obtained their full license. While maintaining the same control variables and GDL indicators, model specifications consider different dependent variables according to the calculation of traffic-related fatalities per state and per year for different age cohorts. Specifically, fatalities are counted for teenagers between 15 and 20 years old (i.e., six age cohorts) and for young adults belonging to the 18-20-, 21-23-, and 24-26-year-old groups (i.e., four additional age cohorts). On the one hand, estimates for the oldest age cohort, which should not be influenced by GDL laws since most of its members did not obtain their full license under these policies, control for possible specification biases. On the other hand, estimates for all age groups evaluate whether the GDL effect is limited to the learner and provisional stages, or also extends to the full licensure period.

The remainder of this chapter is structured as follows. Section 2 introduces the rationale of GDL laws and illustrates findings from previous evaluations of GDL effectiveness in the United States at the single and multi-jurisdictional level. Section 3 presents the data used in the analysis by detailing the calculation of fatalities, the construction of GDL indicators and the definition of control variables. Section 4 illustrates the methodology applied for the econometric evaluations by describing the negative binomial model with fixed effects. Section 5 shows the results and discusses the major outcomes in terms of GDL effectiveness. Section 6 draws the conclusions in terms of model estimates, value of statistical lives and possible policy implications.

2. Graduated Driver Licensing

This section describes the rationale behind the enactment of GDL regulations and the taxonomy proposed by the IIHS, and illustrates findings from previous evaluation of GDL effectiveness at the single and multi-jurisdictional level.

2.1. Rationale of GDL Programs and IIHS Taxonomy

The rationale behind graduated licensing is to delay full licensure, while allowing young novice drivers to obtain their initial experience under lower-risk conditions. GDL programs consist of three stages: a minimum supervised learner's period, an intermediate license (upon passing the driving test) that limits unsupervised driving in high-risk situations, and a full-privilege license available after completion of the first two stages. Young novice drivers must remain in each of the first two stages for predetermined minimum periods.

The IIHS (2008) defines the optimal GDL system as follows: (i) 16 is the minimum age for the learner's permit; (ii) 6 months is the minimum length of the learner stage, during which 30-50 hours of supervised driving are required to be certified by the parents of the young driver; (iii) 18 is the minimum age for the conclusion of the intermediate stage, during which 9pm or 10pm is the latest permitted hour for night driving, and one is the maximum number of teenage passengers allowed. No state in the United States has an optimal graduated licensing system according to this definition. Moreover, different states have approved systems that are significantly different not only according to the number and strength of their components, but also according to their enforcement. In fact, not only are several key components of GDL policies objectively difficult to monitor and enforce (e.g., mandatory supervised driving hours, night driving limitations, teen passenger restrictions), but also some states implement only secondary enforcement that prohibits police from stopping young drivers solely for violating night driving or teen passenger restrictions.

Given the objective difficulties in defining enforcement differences, the IIHS (2008) classifies GDL programs on the basis of their key components, such as the length of the learner's period, the duration of the intermediate phase and the stringency of its restrictions. Accordingly, this taxonomy depends on the attribution of points to the key components of graduated licensing systems, following the criteria illustrated in Table 1. "Good" systems score 6 or more points, "fair" systems score 4 or 5, "marginal" systems 2 or 3, and "poor" ones score less than 2 points (IIHS, 2008). Regardless of point totals, this taxonomy does not rate any state above "marginal" if intermediate license holders can be younger than 16, or if unrestricted driving is allowed before 16 years and 6 months of age. Moreover, this taxonomy characterizes as "marginal" the programs of some states that include an intermediate phase in their GDL regulation, and categorizes as "poor" the programs of other states that do not include an intermediate phase, even though containing components sufficient to obtain a "marginal" or "fair" classification.

The definition of the temporal scope of the analyses presented in this chapter derives from the necessity of an adequate pre- and post-GDL implementation period for all states, and the availability of relevant information about fatal crashes and licensing regulations. Since previous analyses of GDL effectiveness at the multi-jurisdictional level commented that the statistical power of the inferences could have been limited by the ten-year extent of the study period (e.g., Dee et al., 2005; Morrisey et al., 2006), a fifteen-year period was examined. Since FARS databases contain information about fatal crashes on U.S. public roads through the end of 2006, the study period was defined as the years between 1992 and 2006.

Given the defined study period, Table 2 summarizes the effective dates of GDL laws enacted in 48 U.S. jurisdictions between 1992 and 2006. As in much of the prior literature (e.g., Dee et al., 2005; Morrisey et al., 2006), Alaska, Hawaii and the District of Columbia were excluded from the analyses. Note that Florida was the first jurisdiction to enact a graduated licensing system in 1996. Five years later, 28 states had approved GDL laws, even though only four met the IIHS requisites for a "good" characterization. Ten years later, 44 states had passed GDL regulations, and 18 programs contained sufficient key components for a "good" classification.

Note also that, in the two years following the conclusion of the study period, the four remaining states without graduated licensing systems (Arizona, Kansas, Minnesota and North Dakota) approved and enacted new GDL regulations, while seven other states (Connecticut, Delaware, Illinois, Missouri, Nebraska, Nevada and Ohio) improved some components of existing GDL programs and received a higher IIHS classification. The enactment of amendments to GDL laws indicates the existence of a trend among states that they intend to reduce teen traffic fatalities by promulgating rules that are more restrictive. The specification of negative binomial models with indicators of different GDL programs according to different classes within the IIHS taxonomy allows the evaluation of the appropriateness of this trend.

2.2. Studies of GDL Effectiveness for Single Jurisdictions

The majority of the studies about GDL effectiveness has focused on outcomes within single jurisdictions. These studies were intended to evaluate crash rates (e.g., per capita, per licensed driver) for pre-post comparisons, to examine appropriate age groups, to account for long enough follow-up periods, to consider older drivers as a comparison term to control for changes in the general traffic environment, and to adopt adequate statistical techniques to address the evaluation questions. Not all the studies considered all these features, and for this reason some studies require the application of caution in interpreting their results (Shope, 2007).

Table 1. Schedule of points for IIHS taxonomy of licensing systems for young drivers

GDL component	Points assigned
Learner's entry age	1 point for learner's entry age of 16
Learner's permit holding period	2 points for more than 6 months 1 point for 3 to 5 months 0 points for less than 3 months
Practice driving certification	1 point for more than 30 hours 0 points for less than 30 hours
Night driving restriction	2 points for starting at 9pm or 10pm 1 point for starting after 10pm
Teen passenger restriction	2 points for one or less teen passengers 1 point for two teen passengers 0 points for three or more teen passengers
Driver education	where completion of driver education changed a requirement, point values were determined for the driver education track
Duration of restrictions	1 point if the difference between minimum unrestricted license age and minimum intermediate license age is 12 or more months independent evaluation of night driving and teen passenger restrictions

Source: Insurance Institute for Highway Safety

Table 2. Effective dates of graduated licensing laws: 1992-2006

State	GDL laws	IIHS taxonomy of GDL laws		
		Marginal	Fair	Good
Alabama	October 2002		October 2002	
Arizona				
Arkansas	July 2002	July 2002		
California	July 1998			July 1998
Colorado	July 1999		July 1999	July 2005
Connecticut	October 2003		October 2003	
Delaware	July 1999		July 1999	
Florida	July 1996		July 1996	
Georgia	July 1997		July 1997	January 2002
Idaho	January 2001	January 2001		
Illinois	January 1998		January 1998	
Indiana	January 1999		January 1999	
Iowa	January 1999		January 1999	
Kansas				
Kentucky	October 2006			October 2006
Louisiana	January 1998		January 1998	
Maine	August 2000	August 2000		September 2003
Maryland	July 1999		July 1999	October 2005
Massachusetts	November 1998			November 1998
Michigan	April 1997		April 1997	
Minnesota				
Mississippi	July 2000	July 2000		
Missouri	January 2001		January 2001	
Montana	July 2006	July 2006		
Nebraska	January 1999	January 1999		
Nevada	July 2001	July 2001		
New Hampshire	January 1998		January 1998	
New Jersey	January 2001			January 2001
New Mexico	January 2000	January 2000		
New York	September 2003			September 2003
North Carolina	December 1997		December 1997	December 2002
North Dakota				
Ohio	January 1999		January 1999	
Oklahoma	May 1999	May 1999		November 2005
Oregon	March 2000			March 2000
Pennsylvania	December 1999			December 1999
Rhode Island	January 1999		January 1999	July 2003
South Carolina	July 1998	July 1998		
South Dakota	January 1999	January 1999		
Tennessee	July 2001			July 2001
Texas	January 2002		January 2002	
Utah	July 1999		July 1999	August 2006
Vermont	July 2000		July 2000	
Virginia	July 2001		July 2001	
Washington	July 2001			July 2001
West Virginia	January 2001		January 2001	
Wisconsin	September 2000			September 2000
Wyoming	September 2005		September 2005	

Despite differences in temporal scopes, crash data examined, research methods applied and control variables considered, all these studies agree that each unique GDL regulation reduces teenage crash rates, fatalities and hospitalizations to some extent. The examined GDL programs were enacted in U.S. jurisdictions between 1997 and 2002, and classified by IIHS as either "good" (California, Georgia, North Carolina, Pennsylvania and Wisconsin) or "fair" (Iowa, Maryland, Michigan, Texas and Utah). Study periods varied from as early as 1991 to as late as 2005 and included pre- and post-GDL implementation periods lasting from one to seven years. Data consisted of state or provincial records of crashes, as well as census or licensing data to calculate rates. Studies applied a large variety of research methods, which included simple counts, descriptive statistics, crash rates, relative risks, odds ratio, regressions, stochastic multiple structural models, difference-in-differences, and time series analyses. Control variables focused often on specific components of graduated licensing programs, such as restrictions on night driving and teen passengers, and seldom on other traffic legislation, such as zero tolerance laws or seatbelt enforcement methods.

Outcomes from the implementation of GDL programs were evaluated among young novice drivers of different ages, mainly 16- and 17-year-olds. Only one study, which dealt with the implementation of the Iowa licensing system, examined GDL effects on crash rates among 14- and 15-year-old drivers, and obtained a 43% and a 15% reduction, respectively (Hallmark et al., 2008). Only two studies evaluated GDL outcomes on crash rates among 18- and 19-year-olds in North Carolina and California, and only one reported a change in terms of a 17% reduction of fatalities among 18-year-olds in California (Males, 2006).

The effect of GDL regulations for 16-year-olds was evaluated in terms of crash rate reduction as equal to 5% in Utah (Hyde et al., 2005), 14% in Wisconsin (Fohr et al., 2005), 18% in Maryland (Kirley et al., 2008), 23% in California (Rice et al., 2004), 28% in Pennsylvania (Coben and McKay, 2000), 29% in Michigan (Shope & Molnar, 2004), 30% in Iowa (Hallmark et al., 2008), and 57% in North Carolina (Foss et al., 2001). Higher values were obtained in terms of crash fatality reduction, with a 22% reduction estimated in Texas (Willis, 2006), 24% in California (Males, 2006), 30% in Georgia (Rios et al., 2006), 44% in Michigan (Shope & Molnar, 2004), and 49% in Pennsylvania (Coben and McKay, 2000). Notably, a different study estimated no variation as a result of the implementation of the California GDL law (Masten and Hagge, 2004).

The outcome of GDL programs on 17-year-olds was evaluated in terms of crash rate reduction as equal to 6% in Wisconsin (Fohr et al., 2005), 13% in California (Rice et al., 2004), 20% in North Carolina (Margolis et al., 2007), and 25% in Iowa (Hallmark et al., 2008). Inconclusive results were obtained in terms of crash fatalities, with no reduction reported in California (Males, 2006) and a 19% reduction estimated in Georgia (Rios et al., 2006).

If the question concerning the effectiveness of graduated licensing systems is limited to whether or not there such programs are effective, the answer is clearly positive. If the question involves the extent of the outcomes of GDL implementation, the variety of the aforementioned results leads to an inconclusive answer. For 16-year-olds, for example, there is a wide rage in the magnitude of the effect presented, from the 5% crash rate reduction estimated in Utah (Hyde et al., 2005) to the 57% crash rate reduction evaluated in North Carolina (Foss et al., 2001).

Several reasons may explain this large variation in the estimates of GDL effects. First, the stringency of the licensing policies could directly affect their estimated outcomes. However, results rule out this conjecture, as the "good" Wisconsin program appears to be far less effective than the "fair" Iowa program. Second, the analytical methods applied in different studies could affect evaluations of GDL efficacy. In fact, results confirm this supposition; for example, according to Masten and Hagge (2004), the California GDL law apparently does not reduce fatalities among 16-year-olds, while according to Rice et al. (2004) and Males (2006), the same law reduces them by 23% and 24%, respectively. Third, temporal scopes of different studies could influence GDL outcomes. In fact, results substantiate this postulation, as apparently considering longer pre- and post-GDL periods shows some proportionality with the crash rate reduction. Last, the omission of other traffic laws enacted within states during the study period could amplify estimates of GDL effectiveness. Almost all the studies neglected other potential determinants of fatality reduction (such as zero tolerance, seatbelt and speed limit regulations), even though alcohol, failure to wear seatbelts and speeding are common risky behaviors among teenagers. Most relevantly, studies considering seatbelt laws (Hyde et al., 2005; Willis, 2006) report lower estimates of crash rates and fatalities reduction as a result of GDL implementation.

2.3. Studies of GDL Effectiveness for Multiple Jurisdictions

A limited number of studies dealing with GDL effectiveness have covered the United States as a whole. As was observed in studies concentrating on single states, analyses focusing on multiple jurisdictions also revealed differences existing with respect to the age cohorts examined, study goals, data retrieved, temporal scopes considered, research methods applied and control variables included.

Among the six existing studies, one concentrated on 16-20-year-olds, two focused on 15-17-year-olds, and three focused only on 16-year-olds. The outcomes of all the studies were either the number of traffic fatalities or the involvement in fatal or injury crashes. All the studies utilized fatality data from the FARS databases, and some studies used either U.S. Census Bureau data to account for exposure to risk or Federal Highway Administration licensing data to calculate rates. Study intervals consisted of ten-year periods that covered the significant growth in the number of states implementing GDL systems, and yet included states without GDL programs for comparison. Studies applied a large variety of analytical methods, which included chi square for crash rate differences, Poisson regression for adjusted trends estimation, adjusted incidence rate ratios, weighted least squares regression, negative binomial regressions, and conditional negative binomial regressions with fixed effects.

Study goals differed among studies. Eisenberg (2003) did not detail the approach to GDL evaluation, as licensing policies were only one of several measures considered to affect fatal crashes. Williams et al. (2005) underlined trends associated with increasing implementation of GDL regulations. Chen et al. (2006) and Baker et al. (2007) compared jurisdictions with quarters of years under GDL with those not under GDL in three approaches: having three-stage GDL or not, having a specific number of GDL components, or having specific combinations of GDL components not related to the age of licensure. Dee et al. (2005) and Morrisey et al. (2006) used the IIHS classification to assess the effectiveness of "good, fair, marginal and poor" programs on the reduction of teen traffic-related fatalities.

All the studies reported some significant reductions in traffic fatalities or fatal crash involvement. Dee et al. (2005) assessed a 5.6% reduction in traffic fatalities among 15-17-

year-olds, but also showed a 19% reduction for "good" GDL programs. Several studies confirmed that higher reductions were associated with highly rated GDL regulations (Chen et al., 2006; Baker et al., 2007), and Morrisey et al. (2006) also illustrated a 6% and 5% reduction in traffic fatalities among 15-17-year-olds following the enactment of "fair" and "marginal" programs, respectively. Eisenberg (2003) estimated a 9% reduction in fatal crash involvement among 16-20-year-old drivers. Williams et al. (2005) evaluated a 13% reduction in fatalities among 16-year-old drivers. Chen et al. (2006) and Baker et al. (2007) found significantly higher reductions when comparing state-by-quarter observations for GDL programs in which five components were implemented against observations where GDL laws did not implement such components. The reduction in fatal and injury crash involvement among 16-year-olds was estimated equal to 38% and 40%, respectively.

As for single-jurisdictional studies, the question of whether GDL systems have an effect receives a clear, positive answer, while the question concerning the extent of the effect gets an inconclusive answer. First, more stringent GDL programs appear more effective in terms of fatality reduction, corroborating an assumption previously rejected while observing single jurisdiction studies. Second, higher effectiveness of graduated licensing systems was estimated for fatalities and crash involvement among 16-year-olds, rather than among older teenagers. Last, extremely positive results were found for studies that did not take traffic laws (such as zero tolerance, seatbelt and speed limit regulations) into account. Interestingly, minimum reduction values were obtained when considering drunk driving policies (Dee et al., 2005; Eisenberg, 2003), while maximum reduction values were observed when accounting for graduated licensing systems as sole determinants of fatality reduction (Chen et al., 2006; Baker et al., 2007).

3. Data

The analyses presented in this chapter focus on a panel of annual state-level data from 1992 to 2006. Traffic-related fatalities were retrieved from the FARS databases (NHTSA, 2008) for various age cohorts. Crashes included in this census are characterized by the involvement of a motor vehicle traveling on a roadway customarily open to the public and the death of a person (either the occupant of a vehicle or a non-motorist) within 30 days of the crash. Fatalities were counted per state and per year for 15-20-year-olds (i.e., six age cohorts), as well as for teenagers aged 15-17, 18-20, 21-23, and 24-26 (i.e., four additional age cohorts). Given the interest in the overall impact of graduated licensing policies, fatalities include not only drivers, but also passengers and pedestrians. As in much of the prior literature (e.g. Dee et al., 2005; Morrisey et al., 2006), Alaska, Hawaii and the District of Columbia were excluded from the analyses, resulting in final data comprising information over 15 years from 48 states (720 observations) for the ten age cohorts considered.

The analyses use two alternative approaches to characterize GDL implementation. First, a single dummy variable indicates whether, in a given year, a state implemented a three-stage licensing system that included an intermediate phase before a young novice driver could advance to full licensure. As illustrated in Table 2, 44 states enacted graduated licensing systems during the study period. Second, three dummy variables indicate whether, in a given year, a state implemented either a "good", a "fair" or a "marginal" GDL program according to the IIHS taxonomy. States observed in years when they had poor licensing regulations are the reference category. As illustrated in Table 2, at the conclusion of the study period, 18 programs are classified as "good", 17 as "fair" and only 9 as "marginal". Moreover, two states improved from "marginal" to "good" and six states progressed from "fair" to "good" in the last five years of the study period. For both characterizations, when GDL regulations became effective during a calendar year, the appropriate fractional values are taken into consideration.

The analyses control for risk exposure and economic factors within states during the study period. The natural log of the relevant age-specific population (i.e., for 15-20-year-olds, as well as for 15-17-, 18-20-, 21-23-, and 24-26-year-olds) is treated as a control variable of each state's exposure to risk in a given year. Population size estimates at the state level, divided by age, were retrieved from the United States Census Bureau (USCB, 2008). The unemployment rate is considered as a control variable for macroeconomic factors in analyses of state motor vehicle fatality rates (e.g., Evans and Graham, 1988). Annual state unemployment rates were gathered from the Bureau of Labor Statistics of the United States Department of Labor (USDLBLS, 2008).

Table 3. Descriptive statistics for state-by-year panel data: 1992-2006 (720 observations)

Variable	Mean	St. Dev.
Traffic fatalities, age 15	31.57	29.96
Traffic fatalities, age 16	55.32	51.18
Traffic fatalities, age 17	63.40	59.97
Traffic fatalities, age 18	71.14	72.63
Traffic fatalities, age 19	66.78	68.21
Traffic fatalities, age 20	60.61	62.52
Traffic fatalities, ages 15-17	150.29	139.38
Traffic fatalities, ages 18-20	198.53	202.12
Traffic fatalities, ages 21-23	165.35	170.80
Traffic fatalities, ages 24-26	130.43	137.45
Graduated driver licensing law	0.4050	0.4824
Graduated driver licensing law rated "marginal"	0.1568	0.3569
Graduated driver licensing law rated "fair"	0.2331	0.4157
Graduated driver licensing law rated "good"	0.1078	0.3035
65 mph speed limit	0.5756	0.4844
70 mph speed limit	0.4036	0.4823
Seatbelt law - primary enforcement	0.2838	0.4464
Seatbelt law - secondary enforcement	0.9498	0.2152
Zero tolerance law	0.4473	0.4878
Illegal per se at 0.08 BAC	0.4571	0.4864
Administrative license revocation	0.5572	0.4929
State unemployment rate	0.0478	0.0126

Alaska, Hawaii and District of Columbia omitted

The analyses also control for traffic-related legislative measures varying within states during the fifteen-year period, in order to avoid amplification of the GDL effect as a result of the omission of potentially relevant determinants of fatality reduction. Three dummy variables refer to state laws relating to drunk driving, which have been shown to be important predictors of teen traffic fatalities (Dee, 2001). These variables indicate whether it is illegal per se to drive with a positive blood alcohol concentration (BAC) level, if the driver is not of legal drinking age ("zero tolerance" laws), whether it is illegal per se to drive with a BAC of 0.08, and whether the state is allowed to suspend driving licenses before any court action related to a charge of drunk driving ("administrative license revocation"). Two dummy variables control for mandatory seatbelt laws, which have been shown to reduce motor vehicle fatality rates (e.g., Evans and Graham, 1991; Morrisey and Grabowski, 2005). These variables indicate whether states adopt primary enforcement, which allows the police to directly cite motorists for not wearing a seatbelt, or whether states implement only secondary enforcement, which does not allow the police to cite motorists unless they are pulled over for some other reason. Two additional dummy variables identify those states that have increased their rural interstate speed limit to 65 or 70 miles per hour. Recent empirical evidence proves that higher rural interstate speed limits have increased the motor vehicle fatalities on these roads (e.g., Greenstone, 2002). Table 3 presents means and standard deviations of the described variables.

4. Methodology

The analyses presented in this chapter apply a methodology selected on the basis of the choice of the dependent variable and the selection of the model specification.

The considered dependent variable corresponds to the number of fatalities and not to the number of accidents, since deaths are commonly seen as more salient for policy deliberations. Fatality counts are preferred to fatality or crash rates, since the statistical robustness of the results is increased. In fact, weak statistical power because of the reduction of signal-to-noise ratio could be obtained by employing conventional rates. Moreover, high measurement errors could be produced by using rates calculated with respect to estimates of the population data specific to state, year and age cells between census years.

The selected model specification derives from the necessity to acknowledge the count nature of the underlying traffic-related fatality data, and the desire to accommodate the presence of both state and year fixed effects. Accordingly, a conditional maximum likelihood approach to negative binomial models (Hausman et al., 1984), which models count data while allowing for fixed effects, is applied to evaluate the effectiveness of GDL programs in terms of traffic fatality reduction.

Given a panel of states i $(i=1,2,...,N)$ and years t $(t=1,2,...,T)$, the departure point is a Poisson model conditioned on an unobserved mean λ_{it}. The probability that the number y_{it} of traffic-related fatalities occurred within state i during year t is equal to Y_{it}, conditional on λ_{it}, is calculated as follows (Hausman et al., 1984):

$$P\left(Y_{it}=y_{it}\,|\,\lambda_{it}\right)=\frac{\exp\left(-\lambda_{it}\right)\lambda_{it}^{\,y_{it}}}{\Gamma\left(y_{it}+1\right)} \tag{1}$$

The unobserved mean λ_{it} is Gamma distributed with parameters (γ_{it},δ_i). The distribution parameters are expressed as functions of state-by-year exogenous variables x_{it}, parameters β to be estimated, state fixed effects α_i, and state specific over-dispersion parameters φ_i (Hausman et al., 1984):

$$\gamma_{it} = \exp\left(x'_{it}\,\beta\right) \tag{2}$$

$$\delta_i = \varphi_i / \exp\left(\alpha_i\right) \tag{3}$$

The mean and the variance of the number y_{it} of traffic fatalities within state i during year t, conditional on the exogenous variables x_{it}, are dependent on a state specific scale factor $\theta_i = exp(\alpha_i)/\varphi_i$ and are defined as follows (Hausman et al., 1984):

$$E\left[y_{it} | x_{it}\right] = \gamma_{it}/\delta_i = \exp\left(x'_{it}\,\beta + \alpha_i\right)/\varphi_i \tag{4}$$

$$Var\left[y_{it} | x_{it}\right] = \gamma_{it}/\delta_i^2 = \exp\left(x'_{it}\,\beta + 2\alpha_i\right)/\varphi_i^2 \tag{5}$$

The probability of occurrence of a number y_{it} of traffic fatalities for each state i, conditional on state characteristics X_i and the total number of traffic fatalities within state i over the entire study period, is written as follows (Hausman et al., 1984):

$$P\left(y_{i1}, y_{i2}, \ldots, y_{iT_i} \middle| X_i, \sum_{t=1}^{T} y_{it}\right) = \frac{\Gamma\left[\left(\sum_{t=1}^{T} y_{it}\right) + 1\right]\Gamma\left(\sum_{t=1}^{T} \gamma_{it}\right)}{\Gamma\left(\sum_{t=1}^{T} y_{it} + \sum_{t=1}^{T} \gamma_{it}\right)} \prod_{t=1}^{T} \frac{\Gamma\left(y_{it} + \gamma_{it}\right)}{\Gamma\left(y_{it} + 1\right)\Gamma\left(\gamma_{it}\right)} \tag{6}$$

The parameters β are estimated by conditional maximum likelihood (Hausman et al., 1984):

$$\ln L = \sum_{i=1}^{N} \ln P\left(y_{i1}, y_{i2}, \ldots, y_{iT} \middle| X_i, \sum_{t=1}^{T} y_{it}\right) \tag{7}$$

Allison and Waterman (2002) argue that the described negative binomial model is not a true fixed effect method. Greene (2007) explains that the model proposed by Hausman et al. (1984), being conditionally independent of the fixed effects, finesses the incidental parameters problem and provides consistent estimates of the parameters β. Moreover, Greene (2007) clarifies that the state fixed effects are built into the scaling function and not into the conditional mean, thus allowing a potential second set of fixed effects to be defined in the mean of the negative binomial model (i.e., year fixed effects).

In order to measure the effectiveness of GDL policies in traffic fatality reduction, the mean of the number y_{it} of traffic fatalities within state i during year t is rewritten by detailing the components of the matrix of exogenous variables:

$$E[y_{it}] = \exp\left(\alpha_i + x'_{cv,it}\,\beta_{cv} + x'_{gdl,it}\,\beta_{gdl} + \upsilon_t\right) \tag{8}$$

where $x_{cv,it}$ are the control variables for state i at year t, $x_{gdl,it}$ are indicators of GDL policies implemented by state i during year t, α_i are the state fixed effects, υ_t are the year fixed effects, β_{cv} and β_{gdl} are parameters to be estimated. State fixed effects control for state-specific determinants, for example degrees of law enforcement, infrastructure conditions and weather patterns. Year fixed effects control for unobserved national trends in traffic fatalities, for example changes in federal motor vehicle policies and improvements in automotive and medical technologies.

This two-way fixed effects model implies an identification strategy that compares the changes within states that enacted graduated licensing policies to the changes within states that did not, while controlling for both unobserved state-specific and shared year-specific determinants. The comparison consists in the difference between the difference in fatalities before and after the implementation of GDL laws in states that enacted a version of the law, and the difference in fatalities over the same period in states that did not enact the same version of the law. This "difference-in-differences" constitutes the value of the coefficients of interest in the econometric models. The effect of the implementation of GDL programs is measured as the proportionate change in the expected value of the outcome variable due to a unit change of the GDL indicators (Cameron and Trivedi, 1986):

$$\Delta\left(y \middle| x_{gdl}\right) = \left[\exp\left(\beta_{gdl}\Delta x_{gdl}\right) - 1\right] \times 100 \tag{9}$$

5. Results

Negative binomial models are estimated first for the age cohort comprising 15-17-year-old teenagers, then for each of the six single year age cohorts consisting of drivers from 15 to 20 years old, and last for the three older age cohorts (i.e., 18-20-, 21-23-, and 24-26-year-olds). Model specifications consider the indicators of enactment of GDL laws according to the IIHS classification, the control variables described in the data section, and the state and year fixed effects described in the methodological section.

In order to analyze the effectiveness of GDL regulations in terms of fatality count reduction, the general formulation of the conditional mean of traffic-related fatalities assumes the following form:

$$\begin{aligned} E[y_{it}] = \exp(&\alpha_i + \beta_{gdl} \times gdl_{it} + \beta_{zt} \times zt_{it} + \beta_{bac} \times bac_{it} + \beta_{alr} \times alr_{it} + \\ &+ \beta_{sb1} \times sb1_{it} + \beta_{sb2} \times sb2_{it} + \beta_{sl70} \times sl70_{it} + \beta_{sl65} \times sl65_{it} + \beta_{pop} \times pop_{it} + \\ &+ \beta_{unempl} \times unempl_{it} + \textstyle\sum_t \beta_{yeard,t} \times yeard_{it}) \end{aligned} \tag{10}$$

where, within state i during year t, gdl_{it} indicates the enactment of a GDL program, zt_{it} denotes the ratification of a zero tolerance law, bac_{it} implies the endorsement of an illegal-per-se 0.08 BAC law, alr_{it} refers to the enforcement of administrative license revocation, $sb1_{it}$ and $sb2_{it}$ indicate the presence of primary and secondary seatbelt enforcement, $sl70_{it}$ and $sl65_{it}$ denote the existence of 70 and 65 mph speed limits on rural interstate roads, pop_{it} controls for the logarithm of the population, $unempl_{it}$ controls for the unemployment rate, and $yeard_{it}$ are 14 year fixed effects estimated with respect to the 1992 reference year.

When model specifications consider the IIHS classification of GDL programs, expression (10) is modified as follows:

$$\begin{aligned} E[y_{it}] = \exp(&\alpha_i + \beta_{gdlgood} \times gdlgood_{it} + \beta_{gdlfair} \times gdlfair_{it} + \beta_{gdlmarg} \times gdlmarg_{it} + \\ &+ \beta_{zt} \times zt_{it} + \beta_{bac} \times bac_{it} + \beta_{alr} \times alr_{it} + + \beta_{sb1} \times sb1_{it} + \beta_{sb2} \times sb2_{it} + \beta_{sl70} \times sl70_{it} + \\ &+ \beta_{sl65} \times sl65_{it} + \beta_{pop} \times pop_{it} + \beta_{unempl} \times unempl_{it} + \textstyle\sum_t \beta_{yeard,t} \times yeard_{it}) \end{aligned} \tag{11}$$

where, within state i during year t, $gdlgood_{it}$, $gdlfair_{it}$ and $gdlmarg_{it}$, indicate the enactment of a GDL program classified as “good”, “fair” or “marginal” by the IIHS, respectively.

Table 4 illustrates the main evaluation results for 15-17-year-olds from the difference-in-differences evaluation of negative binomial model estimates.

The first model presents estimates for the sparsest specification, which considers only the indicator of GDL implementation within states during different years, the natural log of the state population of the considered age cohort, and the state and year fixed effects. The statistically significant estimate suggests that GDL regulations reduced fatalities by 15.1% among 15-17-year-olds. The second model illustrates estimates with the same sparse specification, only with three indicators of GDL programs classified as “good”, “fair” and “marginal”. “Poor” programs constitute the omitted reference group. Significant parameter estimates indicate that policies that are more stringent are more effective, as “good” programs decreased fatality counts by 20.4%, “fair” programs by 11.1%, and “marginal” programs by 7.8%.

Note that the estimates of both the first and the second model account for unobserved state- and year-specific determinants of fatality reduction, but do not consider other traffic laws and might neglect the effect of potential determinants of teen traffic safety. Clearly, the third and fourth models illustrate that the omission of alcohol-related laws, seatbelt enforcement methods and speed limit regulations could lead to misleading inferences about the effectiveness of GDL implementation. The significant parameter estimated within the third model indicates that GDL laws reduced fatalities by 11.1% among 15-17-year-olds, a negative variation of 26.5% (i.e., [*(15.1-11.1)/15.1*]) with respect to the first model. Moreover, parameters estimated within the fourth model show that “good” programs reduced fatality numbers by 15.1%, “fair” programs by 7.8% and “marginal” programs by 5.1%. The negative variation with respect to the second model amounts to 26%, 29.7% and 34.6%, respectively, and the estimate for the “marginal” program is actually not statistically significant, even at the 10% level.

Even with the inclusion of additional determinants of fatality reduction, inferences regarding the effectiveness of GDL laws show a statistically significant decline of 11.1% in traffic fatality counts among 15-17-year-olds. Moreover, results show a clear, increasing monotonic trend of GDL effectiveness with respect to the stringency of GDL programs, with “good” programs reducing fatalities roughly two times more than “fair” programs, and three times more than “marginal” programs. With respect to nationwide analyses addressing the same age cohorts (e.g., Dee et al., 2005; Morrisey et al., 2006), these results illustrate a larger effect of the implementation of GDL systems in general, and of “fair” and “marginal” programs in particular. Even though slightly lower with respect to previous findings, the estimate of a 15.1% reduction in fatality counts as a result of the implementation of "good" systems is an important result, when considering that almost 11% of the state-by-year observations had these systems enacted. Note that Dee et al. (2005) expressed caution with their initial finding of 19% fatality reduction after the implementation of "good" programs, since only 3% of their state-by-year observations actually had these "good" GDL systems implemented.

Significant parameter estimates within the fourth model show that a decrease in fatality counts among 15-17-year-olds is attributed to zero tolerance laws by 6.6%, and to primary seatbelt enforcement by 5.6%. Parameters for remaining traffic laws are not found to be significant. Neither secondary enforcement of seatbelt laws, nor the speed limits on rural roads, illegal-per-se BAC or administrative license revocation have an effect on fatality counts. On the one hand, these results are consistent with earlier findings by Dee (2001) and Eisenberg (2003), who underlined that youth alcohol control and seatbelt laws were relevant factors toward decreasing teen traffic fatalities. On the other hand, these results contradict previous nationwide analyses (e.g., Dee et al., 2005; Morrisey et al., 2006), which presented inconclusive results regarding the effectiveness of alcohol measures and seatbelt enforcement, although the authors also explained that their short panel most likely caused the weak statistical power.

Table 4. Negative binomial estimates: 15-17-year-old teen traffic fatalities

	Model 1		Model 2		Model 3		Model 4	
Variable	est.	t.stat.	est.	t.stat.	est.	t.stat.	est.	t.stat.
GDL	-0.164	-6.42			-0.118	-2.99		
GDLGOOD			-0.228	-8.30			-0.164	-3.45
GDLFAIR			-0.117	-5.41			-0.082	-2.39
GDLMARG			-0.082	-2.27			-0.052	-1.22
ZT					-0.077	-2.30	-0.068	-1.71
BAC					-0.001	-0.04	-0.013	-0.29
SL70					0.027	1.26	0.022	0.88
SL65					-0.014	-0.69	-0.012	-0.55
SB1					-0.060	-2.60	-0.058	-2.30
SB2					-0.008	-0.22	-0.007	-0.17
ALR					0.014	0.55	0.009	0.33
POP1517	0.355	8.90	0.356	8.38	0.365	6.78	0.366	6.54
UNEMPL					-0.022	-0.03	-0.031	-0.03
Years	15		15		15		15	
States	48		48		48		48	
Observations	720		720		720		720	
Parameters	2		4		10		12	
lnL	-2797.47		-2796.31		-2783.622		-2782.12	
AIC	7.776		7.779		7.760		7.761	

Estimates based on negative binomial regression models that condition on state fixed effects (Hausman et al., 1984). Estimated, but not reported, fourteen dummy variables (years 1993 through 2006) controlled for year fixed effects.

AIC, Akaike Information Criteria, measured as $AIC = -2lnL/n + 2k/n$, where lnL is the log-likelihood function at the estimated parameters, k is the number of parameters and n is the number of observations.

Tables 5 and 6 present the evaluation results for 15-20-year-olds, in order to uncover possible differences in terms of GDL effectiveness among teenagers of different ages. The presented models have the same specification as the fourth model in Table 4.

Table 5. Negative binomial estimates: 15-, 16- and 17-year-old teen traffic fatalities

	15 year olds		16 year olds		17 year olds	
Variable	est.	t.stat.	est.	t.stat.	est.	t.stat.
GDLGOOD	-0.201	-2.94	-0.258	-3.52	-0.075	-2.29
GDLFAIR	-0.104	-2.03	-0.116	-2.53	-0.055	-2.31
GDLMARG	-0.069	-1.53	-0.072	-1.36	-0.008	-0.23
ZT	-0.065	-1.37	-0.099	-2.01	-0.064	-2.65
BAC	-0.078	-0.33	-0.111	-0.46	-0.129	-0.62
SL70	0.045	0.91	0.032	0.95	0.003	0.12
SL65	-0.069	-1.68	-0.014	-0.47	0.009	0.33
SB1	-0.084	-1.88	-0.061	-1.49	-0.069	-3.38
SB2	-0.075	-0.46	-0.055	-0.37	-0.029	-0.35
ALR	0.009	0.21	0.023	0.64	0.015	0.41
POP15	0.356	3.50	0.389	4.66	0.409	4.75
UNEMPL	-0.022	-0.03	0.052	-0.10	0.064	-0.12
Years	15		15		15	
States	48		48		48	
Observations	720		720		720	
Parameters	12		12		12	
lnL	-2106.57		-2300.23		-2310.49	
AIC	5.885		6.423		6.451	

Estimates based on negative binomial regression models that condition on state fixed effects (Hausman et al., 1984). Estimated, but not reported, fourteen dummy variables (years 1993 through 2006) controlled for year fixed effects.

A common pattern is observed with respect to the fatality count reduction resulting from the implementation of GDL laws belonging to different IIHS classes. For example, "good" programs are extremely effective among 15- and 16-year-olds, with a fatality reduction equal to 18.2% and 22.7%, respectively, but less effective among 17-year-olds, with a fatality reduction equal to 7.2%, and not significantly effective for 18-, 19- and 20-year-olds. Estimates for "fair" programs exhibit the same trend, while estimates for "marginal" programs are generally also not significant even for 15-17-year-olds. Apparently, the benefits of graduated licensing systems are felt on novice drivers during the learner and provisional periods, but not after full licensure is obtained at the age of 18. Results for 16-year-olds show lower estimates of GDL effectiveness than previous nationwide studies (e.g., Chen et al., 2005; Baker et al., 2007), which reported much higher effects, but also omitted other potential determinants of teen fatality reduction.

A different pattern is observed for other traffic laws. The zero tolerance law varies between a 6.2% fatality reduction among 17-year-olds and a 9.4% reduction among 16-year-olds. Also, a lower and not significant estimate is observed among 20-year-olds, most likely because these young adults are closer to the legal age limit for alcohol consumption and consequently less influenced by the alcohol prohibition. The primary seatbelt enforcement fluctuates between a 4.4% fatality reduction among 18-year-olds and an 8% reduction among 15-year-olds. Estimates for the remaining traffic laws are not significant for any age group, while the unemployment variable that controls for macroeconomic factors is significant for 19- and 20-year-olds. Logically, unemployment becomes a relevant factor after reaching the post-high-school years.

Results in Tables 4, 5 and 6 consistently suggest that graduated licensing systems caused large and statistically significant reductions in teen traffic-related fatalities. However, these inferences could still be biased by undiagnosed specification errors. On the one hand, the consistency of the estimates and the increasing monotonicity of the parameters with the stringency of GDL policies suggest the inexistence of specification biases. On the other hand, there are attitudinal, institutional and economic variables that could influence the within-state variation in teen fatality counts, but are inherently difficult to measure and consequently problematical to include in model specifications.

Table 6. Negative binomial estimates: 18-, 19- and 20-year-old teen traffic fatalities

	18 year olds		19 year olds		20 year olds	
Variable	est.	t.stat.	est.	t.stat.	est.	t.stat.
GDLGOOD	-0.023	-1.16	-0.030	-0.72	-0.019	-0.29
GDLFAIR	-0.036	-1.58	-0.021	-0.76	-0.022	-0.33
GDLMARG	-0.016	-0.40	-0.012	-0.42	-0.016	-0.31
ZT	-0.076	-3.61	-0.074	-3.27	-0.030	-1.23
BAC	-0.144	-0.68	-0.089	-0.39	-0.120	-0.54
SL70	0.042	1.30	0.046	1.36	-0.007	-0.15
SL65	0.005	0.20	0.000	0.02	-0.006	-0.13
SB1	-0.045	-1.99	-0.060	-1.86	-0.065	-2.00
SB2	-0.020	-0.29	-0.017	-0.24	-0.025	-0.72
ALR	0.028	0.95	0.035	1.32	0.023	0.42
POP15	0.394	6.11	0.380	5.30	0.008	6.96
UNEMPL	0.490	-1.45	0.735	-1.99	0.825	-2.20
Years	15		15		15	
States	48		48		48	
Observations	720		720		720	
Parameters	12		12		12	
lnL	-2384.80		-2372.25		-2456.65	
AIC	6.658		6.623		6.857	

Estimates based on negative binomial regression models that condition on state fixed effects (Hausman et al., 1984). Estimated, but not reported, fourteen dummy variables (years 1993 through 2006) controlled for year fixed effects.

A general robustness check is presented in Table 7 by exploiting panel data on traffic fatalities for the older age cohorts (i.e., 18-20-, 21-23- and 24-26-year-olds). Since most of the GDL regulations were enacted after 1997, the first age cohorts affected by graduated licensing systems were 16-year-olds in either 1997 or 1998. Consequently, age cohorts of 24 or older by 2006 should be unaffected by GDL programs. If the negative binomial models propose reliable inferences about the relationship between GDL changes and fatality counts among 15-17-year-olds, then statistically insignificant effects should be obtained when similarly specified models are estimated among 24-26-year-olds. If significant links were found between GDL changes and traffic fatalities among 24-26-year-olds, they would most likely indicate that omitted determinants of fatality count reduction are confounded with GDL implementation. Table 7 also reports model estimates based on fatality counts among 18-20-, and 21-23-year-olds, while estimates from the fourth model of Table 4 are reproduced for ease of comparison.

Table 7. Negative binomial estimates: Comparison between age cohorts

	15-17 year olds		18-20 year olds		21-23 year olds		24-26 year olds	
Variable	est.	t.stat.	est.	t.stat.	est.	t.stat.	est.	t.stat.
GDLGOOD	-0.164	-3.45	-0.031	-1.09	-0.012	-0.59	-0.013	-0.37
GDLFAIR	-0.082	-2.39	-0.019	-1.11	-0.007	-0.45	-0.006	-0.33
GDLMARG	-0.052	-1.22	-0.019	-0.71	-0.001	-0.05	-0.005	-0.18
ZT	-0.068	-1.71	-0.051	-1.85	-0.009	-0.51	-0.004	-0.08
BAC	-0.013	-0.29	-0.010	-0.15	-0.019	-1.14	-0.021	-0.93
SL70	0.022	0.88	0.039	1.41	0.027	1.80	0.061	1.98
SL65	-0.012	-0.55	-0.021	-1.05	-0.065	-3.95	-0.093	-2.35
SB1	-0.058	-2.30	-0.050	-1.88	-0.046	-2.32	-0.059	-3.59
SB2	-0.007	-0.17	-0.033	-0.52	-0.027	-0.49	-0.018	-0.38
ALR	0.009	0.33	0.031	1.36	0.042	1.12	0.011	0.62
POP1517	0.366	6.54	0.378	7.47	0.403	7.47	0.409	6.77
UNEMPL	-0.022	-0.03	0.762	-2.15	0.688	-1.78	0.850	-1.91
Years	15		15		15		15	
States	48		48		48		48	
Observations	720		720		720		720	
Parameters	12		12		12		12	
lnL	-2782.12		-2873.79		-2703.91		-2565.59	
AIC	7.761		8.016		7.544		7.160	

Estimates based on negative binomial regression models that condition on state fixed effects (Hausman et al., 1984). Estimated, but not reported, fourteen dummy variables (years 1993 through 2006) controlled for year fixed effects.

All three older age cohorts exhibit small and statistically insignificant effects associated with the implementation of either "good", "fair" or "marginal" GDL programs. Estimated parameters for the remaining laws illustrate that, as expected, the zero tolerance law has no effect on age cohorts not concerned with the alcohol prohibition (i.e., those over 21 years old). Primary seatbelt enforcement reduces fatality counts among all the age cohorts, while speed limitations have an effect on fatalities for young adults between 21 and 26 years old. These results suggest that secondary enforcement of seatbelt laws is not effective for any age group, and that only more experienced drivers are sensitive to speed limit enforcement.

Estimates of GDL efficacy for the oldest cohort strongly support the validity of the inferences for the initial model based on fatality counts among 15-17-year-olds. The hypothesis of actual life-saving benefits being shifted to younger adults is not corroborated by parameter estimates, as apparently GDL policies affect only younger novice drivers during the provisional license period. However, even though the effects are not statistically differentiable from zero, all estimates for GDL indicators are consistently negative and imply the possibility that GDL regulations lead to some life-saving benefit. In comparison, Dee et al. (2005) and Morrisey et al. (2006) found insignificant but positive estimates for the same age cohorts, an even more inconclusive finding with respect to this important aspect of graduated licensing measures. Moreover, results require a cautious interpretation because fully licensed drivers (i.e., those over 18 years old) could have obtained their license either before the implementation of a graduated licensing system, or after relocation from a state with a less stringent GDL program(or without such a program at all). Additional years and FARS data would allow further research to investigate a longer panel with updated information.

6. Conclusion

Graduated licensing systems are intended to provide young novice drivers with driving experience under low-risk conditions. Teenagers initially receive a learner's permit that allows them to drive under the supervision of an experienced driver. Then, they receive a provisional license that allows them to drive independently under restrictions addressing risky situations. Lastly, they receive a full license. Through 2008, all jurisdictions in the United States have adopted these three-stage systems.

This study evaluated the effects of GDL programs on teen traffic-related fatalities over the period between 1992 and 2006. More precisely, this study evaluated the difference between the difference in fatalities before and after the implementation of GDL regulations in states that enacted a version of the law, and the difference in fatalities over the same period in states that did not enact the same version of the law. Negative binomial estimates enabled the answering of the questions proposed in the introduction.

The first answer concerns the fact that GDL programs have been quite effective in reducing traffic fatalities among 15-17-year-olds. Parameter estimates illustrate that the average graduated licensing system caused an 11.1% reduction in fatality counts.

The second answer concerns the fact that policies that are more restrictive have been far more effective than less stringent policies. Model estimates show that "good" programs reduced teen traffic fatalities by 15.1%, "fair" programs by 7.8%, and "marginal" programs by 5.1% (even though statistically not significant). Clearly, programs that impose minimum time limits at the learner stage, restrictive driving hours and teen passenger limits at the intermediate stage, and a minimum age for obtaining full licensure, obtain more significant benefits than programs lacking some of these features. Considering the evolution of graduated licensing systems in the last two years (i.e., 2007 and 2008), the trend of policy-makers in several states, who decided to design GDL laws where these were lacking, or to modify existing GDL programs in favor of more restrictive ones, appears to be moving in the correct direction to further reduce teen traffic-related fatalities.

The third answer concerns the fact that GDL regulations appear not to have traffic-safety implications for teenagers who have already reached full licensure. Estimates show that older age cohorts are not significantly affected by graduated licensing policies. Moreover, estimates show that, already at the age of 17, the GDL effect is much inferior than at the age 15 and 16, suggesting that the first year after licensure is strongly affected by licensing policies, while the following one is positively affected, but to a lesser extent. On the one hand, for the older age cohort this result is desirable and assures the absence of specification bias in the models, as almost all the young adults over 24 years old did not receive their license under the policies examined. On the other hand, for 18-23-year-olds, this result is less desirable since it suggests that GDL implementation only influences drivers during their learner's permit and provisional license periods. Foss (2007) correctly comments that the critical issue is to comprehend the larger conceptual goal, which is both to protect young novice drivers while they are learning and to create safer drivers overall. Foss (2007) also explains that results do not illustrate whether GDL programs are successful in creating safer drivers, but clearly show that GDL policies are extremely successful in the task of saving lives during the most critical period, namely the first year after licensure.

From the benefit perspective, model estimates may translate the efficacy of GDL regulations in young lives saved through approximate calculations. In 2006, within the 18 states implementing "good" GDL programs, there were 2,706 traffic fatalities among 15-17-year-olds. A 15.1% fatality reduction implies that there would have been 482 additional fatalities in these states without the implementation of "good" licensing policies. Similar calculations imply that there would also have been 216 more victims in the 17 states adopting "fair" programs, and 36 more deaths in the 9 states enacting "marginal" programs. Moreover, the four states that had not introduced GDL by the end of 2006 had 405 traffic fatalities among 15-17-year-old teenagers. Estimated GDL effects imply that graduated licensing policies would have prevented 21 fatalities because of the approval of "marginal" programs, 32 as a result of the adoption of "fair" programs, and 61 as a consequence of the implementation of "good" programs. If the value of a statistical life for prime-age workers has a median value of about \$7.3 million (Viscusi and Aldy, 2003), the 795 lives saved in states with GDL programs correspond to \$5.3 billion. If the four states without regulations had adopted "good" GDL programs before the end of 2006, the value of the lives saved as a result thereof would have been approximately \$445 million.

From the cost perspective, graduated licensing policies require an additional visit to the license examiner and limited additional law enforcement activities on the part of the states. The most difficult cost to evaluate is most likely the travel disutility generated by the restrictions on night driving and teen passengers. When considering the value of statistical lives in the four states without a GDL program (\$445 million), in relation to their population aged between 15 and 17 years old (approximately 640 thousand teenagers), the dollar benefit per constrained teen is slightly less than \$700. Clearly, there are costs generated by GDL policies, but policy makers and public opinion would most likely overlook such costs in favor of the more relevant number of human lives saved.

Summarizing, model estimates clearly suggest that GDL regulations have been highly effective in reducing the leading cause of fatalities among teenagers, especially where stringent restrictions have been imposed. The enhancement of the structure of GDL systems whenever possible, the improvement of the functioning of existing elements through primary enforcement, and the involvement of parents in educating, coaching or demanding changes in young drivers' behavior, are surely the directions in which jurisdictions should be moving in order to save additional human lives.

REFERENCES

Allison, P. D. & Waterman, R. (2002). Fixed effects negative binomial regression models. *Sociological Methodology, 32 (1)*, 247-265.

Baker, S., Chen, L. & Li, G. (2007). Nationwide review of graduated driver licensing. AAA Foundation for Traffic Safety, Washington, D.C. Available at: http://www.aaafoundation.org/pdf/nationwidereviewofgdl.pdf.

Cameron, A. C. & Trivedi, P. K. (1986). Econometric models based on count data: comparisons and applications of some estimators and tests. *Journal of Applied Econometrics, 1 (1)*, 29-53.

Chen, L., Baker, S. P., Braver, E. R. & Li, G. (2000). Carrying passengers as a risk factor for crashes fatal to 16- and 17-year-old drivers. *Journal of the American Medical Association, 283*, 1578-1617.

Chen, L., Baker, S. P. & Li, G. (2006). Graduated driver licensing programs and fatal crashes of 16-year-old drivers: A national evaluation. *Pediatrics, 118 (1)*, 56-62.

Coben, J. H. & McKay, M. P. (2003). Evaluation of the effectiveness of Pennsylvania's graduated driver licensing program. Commonwealth of Pennsylvania Department of Transportation, Harrisburg, PA.

Dee, T. S. 2001. Does setting limits save lives? The case of 0.08 BAC laws. *Journal of Policy Analysis and Management, 20 (1)*, 111-118.

Dee, T. S., Grabowski, D. C. & Morrisey, M. A. (2005). Graduated driver licensing and teen traffic fatalities. *Journal of Health Economics, 24*, 571-589.

Eisenberg, D. (2003). Evaluating the effectiveness of policies related to drunk driving. *Journal of Policy Analysis and Management, 22 (2)*, 249-273.

Evans, W. N. & Graham, J. D. (1988). Traffic safety and the business cycle. *Alcohol Drugs and Driving, 4*, 31-38.

Evans, W. N. & Graham, J. D. (1991). Risk reduction or risk compensation? The case of mandatory safety-belt use laws. *Journal of Risk and Uncertainty, 4 (1)*, 61-73.

Ferguson, S. A. (2003). Other high risk factors for young drivers: how graduated licensing does, doesn't, or could address them. *Journal of Safety Research, 34 (1)*, 71-77.

Fohr, S. A., Layde, P. M. & Guse, C. E. (2005). Graduated driver licensing in Wisconsin: Does it create safer drivers? *Wisconsin Medical Journal, 104 (7)*, 31-36.

Foss, R. D. (2007). Improving graduated driver licensing systems: A conceptual approach and its implications. *Journal of Safety Research, 38* (), 185-192.

Foss, R. D., Feaganes, J. R. & Rodgman, E. A. (2001). Initial efforts of graduated driver licensing on 16-year-old driver crashes in North Carolina. *Journal of the American Medical Association, 286 (13)*, 1588-1592.

Greene, W. (2007). Functional form and heterogeneity in models for count data. *Foundations and Trends in Econometrics, 1 (2),* 113-218.

Greenstone, M. (2002). A reexamination of resource allocation responses to the 65-MPH speed limit. *Economic Inquiry, 40 (2),* 271-278.

Hallmark, S. L., Veneziano, D. A., Falb, S., Pawlovich, M. & Witt, D. (2008). Evaluation of Iowa's graduated driver's licensing program. *Accident Analysis and Prevention, 40 (4),* 1401-1405.

Hausman, J., Hall, B. H. & Griliches, Z. (1984). Econometric models for count data with an application to the patents-R&D relationship. *Econometrica, 52 (4),* 909-938.

Hyde, L. K., Cook, L. J., Knight, S. & Olson, L. M. (2005). Graduated driver licensing in Utah: Is it effective? *Annals of Emergency Medicine, 45 (2),* 147-154.

Kirley, B. B., Feller, A., Braver, E. & Langenberg, P. (2008). Does the Maryland graduated driver licensing law affect both 16-year-old drivers and those who share the road with them? *Journal of Safety Research, 39 (3),* 295-301.

Insurance Institute for Highway Safety (2006). U.S. licensing systems for young drivers. Insurance Institute for Highway Safety, Arlington, VA. Available at: www.iihs.org/laws/state_laws/pdf/us_licensing_systems.pdf.

Lin, M. L. & Fearn, K. T. (2003). The provisional license: Nighttime and passenger restrictions - a literature review. *Journal of Safety Research, 34 (1),* 51-62.

Males, M. (2006). California's graduated driver license law: Effects on older teenagers. *Californian Journal of Health Promotion, 4 (3),* 207-221.

Margolis, L. H., Masten, S. V. & Foss, R. (2007). The effects of graduated driver licensing on hospitalization rates and charges for 16-and 17-year-olds in North Carolina. *Traffic Injury Prevention, 8* (1), 35-38.

Masten, S. V. & Hagge, R. A. (2004). Evaluation of California's graduated driver licensing program. *Journal of Safety Research, 35 (5),* 523-535.

Mayhew, D. (2003). The learner's permit. *Journal of Safety Research, 34 (1),* 35-44.

Morrisey, M. A. & Grabowski, D. C. (2005). State motor vehicle laws and older drivers. *Health Economics, 14 (4),* 407-419.

Morrisey, M. A., Grabowski, D. C., Dee, T. S. & Campbell, C. (2006). The strength of graduated drivers license programs and fatalities among teen drivers and passengers. *Accident Analysis and Prevention, 38 (1),* 135-141.

National Highway Traffic Safety Administration (2008). Fatal Accident Report Systems. Available at: ftp://ftp.nhtsa.dot.gov/FARS.

Preusser, D. F., Ferguson, S. A. & Williams, A. F. (1998). The effect of teenage passengers on the fatal crash risk of teenage drivers. *Accident Analysis and Prevention, 30 (2),* 217-222.

Preusser, D. F. & Leaf, W. A. (2003). Provisional license. *Journal of Safety Research, 34 (1),* 45-49.

Rice, T. M., Peek-Asa, C. & Kraus, J. F. (2004). Effects of the California graduated driver licensing program. *Journal of Safety Research, 35 (4),* 375-381.

Rios, A., Wald, M., Nelson, S. R., Dark, K. J., Price, M. E. & Kellermann, A. L. (2006). Impact of Georgia's teenage and adult driver responsibility act. *Annals of Emergency Medicine, 47 (4),* 361–369.

Shope, J. T. (2007). Graduated driver licensing: Review of evaluation results since 2002. *Journal of Safety Research, 38 (2),* 165-175.

Shope, J. T. & Molnar, L. J. (2004). Michigan's graduated driver licensing program: Evaluation of the first four years. *Journal of Safety Research, 35 (3),* 337-344.

Simons-Morton, B. G. & Hartos, J. L. (2003). How well do parents manage young drivers? *Journal of Safety Research, 34* (1), 91-97.

United States Census Bureau (2008). Population estimates. Available at: http://www.census.gov/popest/estimates.php.

United States Department of Labor, Bureau of Labor Statistics (2008). Labor statistics data. Available at: http://data.bls.gov/PDQ/outside.jsp?survey=la.

Viscusi, W. K. & Aldy, J. E. (2003). The value of a statistical life: A critical review of market estimates from around the world. National Bureau of Economic Research Working Paper No. 9487.

Williams, A. F. (2003). Patterns of risk. *Journal of Safety Research, 34 (1),* 5-15.

Williams, A. F. (2007). Contribution of the components of graduated licensing to crash reductions. *Journal of Safety Research, 38 (2)* 177-184.

Williams, A. F., Ferguson, S. A. & Wells, J. K. (2005). Sixteen-year-old drivers in fatal crashes, United States, 2003. *Traffic Injury Prevention, 6 (3),* 202-206.

Willis, D. K. (2006). Fatal crashes involving 16 year-old Texas drivers pre-and post-GDL: Who, when, where, and why? Center for Transportation Safety, Texas Transportation Institute, Texas A&M University System, College Station, TX.

In: Accidents: Causes, Analysis and Prevention
Editors: Henri Bédard and Geraud Delashmit
ISBN: 978-1-60741-712-5

Chapter 2

PSYCHOLOGICAL FUNCTIONS OF DRIVING AND RISKY DRIVING AND INVOLVEMENT IN RISKY DRIVING IN ADOLESCENCE

***Manuela Bina*[1]*, Federica Graziano*[*1]*, Emanuela Calandri*[1]*, Giovanni Vecchio*[2] *and Elena Cattelino*[3]**

[1]Department of Psychology, University of Turin, Turin, Italy
[2]University of Rome "La Sapienza", Rome, Italy
[3]University of Aosta Valley, Aosta, Italy

ABSTRACT

Driving a motor vehicle is a key behavior in the development of adolescents; however, driving may be dangerous for teenagers, especially when they fail to respect the traffic code or drive in a risky manner. Risky driving is, in fact, one of the main causes of motor vehicle accidents. Psychological literature has stressed that driving and risky driving in adolescence represent self-regulated actions, oriented towards the attainment of meaningful goals for individual development. In other words, driving and risky driving fulfill psychological functions in relation to adolescent developmental tasks (Jessor, 1998; Bonino et al., 2005). The study of these psychological functions is crucial to understand why adolescents engage in risky driving.

The aims of the study were to investigate: 1) the psychological functions of driving and risky driving as perceived by adolescents; 2) the factors underlying psychological functions of driving and risky driving; 3) the relationships between driving and risky driving functions and involvement in risky driving. With reference to previous studies (Bonino, et al. 2005), two factors underlying the functions of risky driving were expected, the first includes functions related to identity definition (experimentation and perception of control, transgression, sensation seeking, coping) and the second includes functions related to peer relationships (visibility, peer emulation, surpassing peers, identification with peers). Moreover, different levels of involvement in risky driving were expected in relation to different perceived functions.

* Corresponding author: Federica Graziano fgrazian@psych.unito.it

Data were collected through an anonymous self-report questionnaire. The sample included 626 adolescents ages 15 to 19; it was balanced for gender and representative of adolescents attending high schools in Turin (northwestern Italy). Risky driving was evaluated using the scale by Bonino et al. (2005), while scales of psychological functions were defined on the basis of a preliminary explorative study using focus group interviews with adolescent drivers (Bina et al., 2008). Confirmatory factor analysis identified the factors underlying the psychological functions of driving and risky driving. Hierarchical multiple regression analysis investigated the relationships between functions and involvement in risky driving.

Factor analysis revealed two factors underlying driving functions: leisure (driving for fun, to meet friends, and to pass the time) and autonomy (in travel, from parents, to reach distant places) and two factors underlying risky driving functions (identity definition and peer relationships). Regression analysis showed that only identity definition functions were linked to involvement in risky driving. Driving functions (leisure and autonomy) and relationship functions of risky driving were not statistically significant in the regression model.

Implications for risky driving and motor vehicle accident prevention programs are discussed.

INTRODUCTION

Motor vehicle driving is common among adolescents and young people. Nonetheless, driving a car or riding a moped or motorcycle may be considered a risk behavior when teenagers drive in a risky manner or fail to respect the traffic code (in particular, failing to respect speed limits and use safety devices, or driving while intoxicated) (Jessor, 1987, 1998; Shope, Raghunathan & Patil, 2003; Asbridge, Poulin & Donato, 2005; Bonino, Cattelino & Ciairano, 2005). Young drivers' involvement in road traffic accidents is a widely recognized problem: in Italy and in other European countries, motor vehicle accidents are, in fact, the main cause of death among 15- to 34-year-olds (ISTAT/ACI, 2007; ERSO, 2008).

Developmental psychology literature has stressed that driving in adolescence not only allows for greater independence and freedom of movement, but is also a crucial behavior in the processes of identity development and redefinition of relationships with peers and adults (Jessor, 1987, 1998; Bonino et al., 2005). The phenomenon of risky driving can only really be understood in relation to the more general adolescent lifestyle and with reference to a complex system comprised of individual values and attitudes (Laapotti, Keskinen, Hatakka & Katila, 2001; Hatakka, Keskinen, Gregersen, Glad & Hernetkoski, 2002), as well as contextual variables related, in particular, to family, peer, and school contexts (Jessor, 1987; Jessor, Turbin & Costa, 1997). With reference to these theoretical models, many researchers have stressed that driving and risky driving in adolescence, like other risk behaviors, represent self-regulated actions, oriented towards the attainment of meaningful goals for individual development (Silbereisen e Noack, 1998; Costa, Jessor & Turbin, 2007). In other words, driving and risky driving fulfill psychological functions tied to adolescent developmental tasks (Jessor, 1998; Bonino et al., 2005).

Before considering the literature on the psychological functions of risk behaviors, a preliminary statement should be made with reference to the terminology used by the researchers. The concept of "functionality" of risk behaviors was first described by Jessor and colleagues (1977, 1991). The term was also used by Silbereisen, Eyferth & Rudinger (1986), who stressed the concept of "functional equivalence" of behavior. In other words, not only can the same behavior fulfill different psychological functions for the individual who engages in it (e.g. smoking cigarettes as a way to experience something new, or as a way to emulate peers), but different behaviors can also originate from the same function (e.g. showing autonomy by smoking or by sustaining an individual point of view during a discussion). To our knowledge, studies referring strictly to the concept of function are not very numerous. However, other research has investigated concepts very similar to functions using different terminology. Spruijt-Metz (1999), Spruijt-Metz, Gallagher, Unger & Anderson-Johnson (2004), and Spruijt-Metz, Gallagher, Unger & Johnson (2005) studied *meanings* of risk behaviors, while other researchers have focused on *justifications* (Shapiro, Siegel, Scovill & Hays, 1998). Still other studies have dealt with *motives* (Newcomb, Chou, Bentler & Huba, 1988; Cooper, 1994; Cooper, Shapiro & Powers, 1998; Gebhardt, Kuyper & Greunswen, 2003; Kloep, Guney, Cok & Simsek, in press) or *reasons* (Robinson, Holmbeck & Paikoff, 2007) for risk behaviors. Despite the lack of common terminology, all these studies have pointed to the fact that risk behaviors are not senseless and are not merely about sensation-seeking or the imitation of peers, but rather that these behaviors can only be understood by investigating the psychological aspects (referred to as functions, motivations, or reasons) salient for the individual who acts in a risky way.

Keeping these aspects in mind, studies concerning psychological functions of risk behaviors in adolescence can be divided into two main groups. The first is represented by studies that have investigated functions indirectly, using quantitative data on large samples of adolescents, and considering, on one hand, adolescent implication in risk behaviors and, on the other hand, the individual and social correlates and predictors of this implication. In this case, the functions have been inferred by looking at the strength of the relationships between these elements (Jessor, Donovan & Costa, 1991; Engels, 1998; Engels & ter Bogt, 2001; Bonino et al., 2005). For instance, the relation between drinking and friend relationships characterized by high levels of perceived support from peers and peer attachment, suggests that this behavior carries out functions connected to socialization and to the sense of belonging to a group (Engels & ter Bogt, 2001). In a study by Bonino et al. (2005), the link between transgression and heavy drinking was suggested by the relationships between drinking and behaviors characterized by infraction of laws, such as deviance and risky driving, as well as individual variables, such as high approval of deviance, little importance attached to school achievement, and high orientation towards peers.

The second group of studies on functions is represented by research that has investigated directly the psychological functions that adolescents assign to specific risk behaviors. The majority of these studies have used qualitative investigation instruments, such as individual interviews or focus groups, in order to let adolescents speak about functions fulfilled by different behaviors. In some cases, research based on qualitative instruments was preliminary to a subsequent investigation using quantitative instruments. In particular, with reference to the Italian context, functions of sexual intercourse and of psychoactive substance use have been investigated (Giannotta, Ciairano, Bonino & Morero, 2004; Giannotta, Ciairano, Spruijt & Spruijt-Metz, in press), including a cross-national comparison with Netherlands (Ciairano, 2004). Other European studies have focused on psychosocial functions of risky driving (Møller, 2004; Møller and Gregersen, 2008), on motives for having sex (Gebhardt et al., 2003), and on motives for risk taking (Kloep et al., in press). In the North-American context, similar studies have explored the meanings of cigarette smoking (Hsia e Spruijt-Metz, 2003;

Spruijt-Metz et al., 2004; Spruijt-Metz et al., 2005; Tilleczek e Hine, 2006), the meanings of unhealthy eating patterns and physical exercise (Spruijt-Metz, 1999), motives for alcohol and other psychoactive substance use (Newcomb et al., 1988; Cooper, 1994), motives or reasons for having sex (Cooper et al, 1998; Robinson et al., 2007), as well as justifications for risk-taking (Shapiro et al., 1998).

In relation to driving and risky driving, the principal studies that inferred psychological functions were carried out by Jessor (1998) and Bonino et al. (2005). Risky driving seems to fulfill a desire to show off, to demonstrate adulthood, to gain the appreciation of one's peer group, to master risk, and to overcome one's own limits (Jessor, 1998). In a study by Bonino et al. (2005), driving behavior seemed to fulfill a need for practical and psychological autonomy, physical and psychological separation from conflict situations, and adulthood. Similar to the findings of Jessor's study, functions attached to risky driving were related to identity definition and social relationships with peers and adults. In particular, functions related to identity were adulthood (risky driving as an exaggerated way to exhibit an adult identity in a superficial manner), experimentation, and perception of control (risky driving as a way to test one's driving skills and to prove one's own fearlessness), transgression (violation of the traffic code), coping strategies and escape (risky driving as an emotion-focused strategy to cope with problems), and sensation-seeking. Functions related to relationships were self-affirmation in front of peers or adults, also by showing off with powerful vehicles, identification with peers and differentiation from adults, peer emulation, and surpassing peers (risky driving as a challenge to measure oneself against peers).

Of the studies that have investigated functions of driving and risky driving directly, some of the most noteworthy research was conducted by Shapiro et al. (1998), Møller (2004) and Møller and Gregersen (2008).

In the study by Shapiro et al. (1998) speeding, drunk driving, and reckless driving were investigated in a wider study of justifications for different risk-taking behaviors. The study was conducted in a sample of college-age adolescent females, using a risk-taking questionnaire and a weekly diary of risk-taking behaviors. Three main categories of justifications were identified: functional (goal-oriented, intentional, and instrumental, e.g. time management justifications), self-centric (psychological, physical, and hedonic justifications) and social (justifications related to promoting, discouraging, or enhancing contact with other people). Infrequent risk-takers reported more functional justifications while frequent risk-takers reported more social and self-centric justifications.

In Møller's (2004) qualitative explorative study, based on focus group interviews with a sample of 18-24-year-old drivers, functions of driving were investigated and relationships between risky driving and lifestyle were studied. Four psychosocial functions of driving were identified: visibility (driving as a way to draw attention), status (driving as a means of expressing status as part of one's self-identity), control (over the vehicle), and mobility (driving as a way to achieve physical and psychological autonomy). Møller stressed the utility of taking into account psychological functions when attempting to explain the relationships between risky driving and lifestyle: in fact, she demonstrated that a lifestyle with few planned activities, few hobbies, and frequent meetings with friends seems to facilitate the use of the car in a way that promotes risk-taking. Starting from these preliminary results, Møller and Gregersen (2008) examined the relationships between risk-taking while driving, the psychosocial function of driving, leisure time activities, car-oriented peer group interactions, and educational attainment in a large sample of 18-25-year-old drivers in Denmark, by using a mailed questionnaire. The function of driving was measured through nine items included in the analysis separately as well as in the total score. Main results stressed that the psychosocial function of driving is an indicator of risk-taking while driving, in particular, status and blowing off steam are linked to a greater involvement in risky driving, while becoming an adult is tied to less risky driving.

Based on the definition offered by Bonino et al. (2005) of functions of driving and risky driving, in a previous study we directly investigated functions of both behaviors by using focus group interviews with a sample of 15-19-year-old Italian adolescent drivers (see Bina, Graziano, Borca & Calandri, 2008 for results concerning functions of risky driving). Interviews were analyzed with content analysis and sentences referring to functions of risky driving were classified in ten categories: the most frequently mentioned functions were self-affirmation, experimentation and perception of control, peer emulation and surpassing peers. The other categories were sensation-seeking, identification with peers, adulthood, coping strategies and escape, sharing of actions and emotions, pleasure and fun, and instrumental uses.

In the present chapter we present the results of a study in which functions of driving and risky driving were investigated directly using a questionnaire developed based on the results of focus group interviews. In particular, specific items about functions of driving and risky driving were formulated and submitted to a sample of Italian adolescents who were also asked to answer questions about their involvement in driving and risky driving. This quantitative study allowed us to have data on a larger sample of adolescents than the qualitative preliminary study, as well as to investigate relationships between perceived functions and involvement in driving and risky driving.

In the framework of literature on functions of risk behaviors, the present study directly investigates the functions of driving and risky driving, defined, respectively, as advantages of driving a motor vehicle and reasons for driving a motor vehicle in a risky manner. While the above-mentioned studies on the functions of risky driving did not distinguish between risky driving and other risk behaviors (Shapiro et al., 1998), nor between driving and risky driving (Møller, 2004; Møller and Gregersen, 2008), this study makes a clear distinction between the two different behaviors. Moreover the study was carried out on a sample of adolescents between the ages of 15 and 19, while other studies mentioned investigated the topic in older populations (18 year of age or older). This allowed us to consider also risky driving and functions related to riding a moped or motorcycle and not only to car driving. Mopeds in particular are a popular means of transportation among adolescents and the problem of traffic accidents involving moped drivers can not be overlooked (Kopjar, 1999; Michel, Purper-Ouakil & Mouren-Siméoni, 2002). From a methodological point of view, the main aim of the present study was to identify the factors underlying the psychological functions of driving and risky driving by carrying out confirmatory factor analysis, whereas other studies on functions have been based on single items (Møller and Gregersen, 2008) or have only employed explorative factor analysis (Kloep et al., in press) or scale definitions based on agreement between coders (Robinson et al., 2007). Moreover the goal of the present study was to investigate relationships between functions and implication in risky driving. In fact, only through the study of the links between the motivational and behavioral aspects of risky driving can we hope to obtain useful indications for the implementation of risky driving and motor vehicle accident prevention programs targeting adolescent drivers.

METHOD

Aims

Based on the theoretical framework outlined, the aims of the study were to investigate:

1) the psychological functions of driving and risky driving perceived by adolescents (also considering differences between drivers and non-drivers);
2) the factorial structure of the psychological function scales of driving and risky driving;
3) the relationships between driving functions, risky driving functions, and risky driving. In particular we intended to analyze the joint effects of driving and risky driving functions on adolescents' involvement in risky driving as well as the specific contribution of each different factor of these functions (considering gender, age, type of school, mileage, and type of vehicle driven).

With regard to our second aim and based on the findings of previous studies (Bonino, et al. 2005), two factors underlying the functions of risky driving were expected: first, functions related to identity definition (experimentation and perception of control, transgression, sensation-seeking, coping); and second, functions related to peer relationships (visibility, peer emulation and surpassing peers, identification with peers). Finally, with reference to the third aim, different levels of involvement in risky driving in relation to different perceived functions were expected.

Participants

The research involved 626 adolescents, boys (45%) and girls (55%), ages 15 to 19 (mean age 17,1; 41% >18 years old, 42% 16-17 years old, 17% 15 years old), living in the province of Turin (northwestern Italy) and attending different high schools (lyceums, technical, and professional schools)[1].

Of the total sample, 317 (51% of participants; 56% boys) were drivers, defined as adolescents who had driven at least one motor vehicle - car, motorbike, moped [2] - over the last two months. The majority of the adolescent drivers had driven a car (53%)[3] or a moped

---

[1] In Italy there are three main paths for secondary school with various specializations: lyceums (classical, scientific, linguistic, psycho-pedagogic, artistic schools), technical schools (e.g. for bookkeepers, industrial technicians, surveyors), and professional schools (e. g. for artisans, tourism operators, social service operators) (Bonino et al., 2006).

[2] By Italian law, mopeds (with piston displacements lower than 125 cc) can be driven from the age of 14 with a specific driving license, motorcycles (with piston displacements higher than 125 cc) can be driven from the age of 16 with a type A driving license, while cars can be driven from the age of 18 with a type B driving license.

[3] 33% of car drivers drove without having a type B driving license and more than 50% of motorbike drivers drove without having a type A driving license. This result is in line with the findings of previous studies (Bonino et al., 2005).

(57%), while 10% had driven a motorcycle[4]. In terms of mileage, without considering the type of vehicle driven, 45% of the participants drove 1 to 10 km per week, 17% 11 to 30 km, 15% 31 to 50 km, 12% 51 to 100 km, and finally 11% more than 100 km.

Instrument

An anonymous self-report questionnaire was used to collect the data. The questionnaire, specifically defined for the aims of the study, investigated aspects of driving behavior, implication in risky driving, and the functions of these behaviors. The scales of psychological functions were developed following a preliminary explorative study based on focus group interviews with adolescent drivers (Bina et al., 2008). The questionnaire was administered by trained researchers in the schools to different classes during class time. Teachers remained outside the classroom while questionnaires were administered. Completed questionnaires were turned in immediately. Data were collected in accordance with Italian law and the ethical code of the Professional Psychologists Association. Measures used for the present study are presented.

Driving functions

To investigate psychological driving functions, the adolescents' level of agreement with eleven items concerning the benefits of driving a motor vehicle (e.g. *to feel more independent from your parents, to feel grown up, to be able to meet friends)* was measured. All the items are reported in Table 1. Adolescents responded on a 4-point Likert scale, ranging from 0 = *disagree* to 3 = *strongly agree* (1 = *partially agree*; 2 = *agree*). All participants (drivers and non-drivers) responded to questions about the functions of driving.

Risky driving functions

To investigate the psychological functions of risky driving we asked adolescents to what extent they agreed with fourteen items concerning the reasons for driving a motor vehicle in a risky manner (like *because it's exciting* or *to test my driving skills,* or *to show off in front of friends*). All the items are reported in Table 5. Adolescents responded on a 4-point Likert scale, ranging from 0 = *disagree* to 3 = *strongly agree* (1 = *partially agree*; 2 = *agree*). Items were introduced with the phrase *"I (would) drive in a risky way because..."* in order to allow all participants (drivers and non-drivers) to respond. This formulation was chosen with reference to previous studies which stressed its reliability in studying the meanings of behaviors (Spruijt-Metz et al., 2004).

Risky driving

The risky driving scale (Bonino et al., 2005) evaluates the frequency with which adolescents report committing twelve different traffic offences (e.g. exceeding the speed

[4] The total exceeds 100 because many participants had driven more than one type of motor vehicle during the last two months.

limit, running a red light, driving too closely to the car in front of you, driving under the effect of alcohol or marijuana) and risky driving behaviors (e.g. taking risks while driving because it makes driving more fun) over the last two months. For each of the 12 items adolescents responded on a 5-point Likert scale, ranging from 0=*never* to 4= *6 or more times* (1= *once*; 2= *2-3 times*; 3= *4-5 times*) (range 0-48; N= 203; mean= 7.89; sd= 9.24; Cronbach alpha= .88).

RESULTS

Driving Functions

With reference to driving functions, the great majority of adolescents attributed various advantages to driving motor vehicles (the percentages of agreement vary between 98% and 32%, calculated by summing *agree* and *strongly agree*). In particular, adolescents perceived the main advantages of driving a car or a motorcycle to be the ability to reach more distant places (98%), to be able to travel more independently (96%), and to feel more independent from parents (83%) (see Table 1).

These results suggest that the main functions of driving concern mobility and independence.

Further, advantages of driving related to leisure time, like to meet friends (80%) and to have fun (80%), were also endorsed by the majority of participants. The function of adulthood was less endorsed by adolescents (to feel grown up 42%).

Some significant differences between drivers and non-drivers were found. In particular, adolescent drivers were more likely than non-drivers to think that driving a motor vehicle could enable them to feel more independent from their parents and could be a fun way of spending time.

Table 1. Driving functions: percentages of agreement (sum of *agree* and *strongly agree*)

Items	% Total	% Drivers	% Non-drivers	χ^2	p
It lets you get to more distant places	98%	96%	99%	4.21	.040
It lets you travel more independently	96%	96%	95%	0.31	.580
It lets you get to places faster when going out with friends	86%	87%	86%	0.27	.601
It makes you feel more responsible	84%	86%	83%	0.75	.386
It makes you feel more independent from your parents	83%	87%	79%	7.89	.005
It gives you a sense of freedom	82%	84%	81%	1.40	.237
It's fun	80%	86%	73%	18.27	.000
It's a good opportunity to meet friends	80%	83%	78%	2.17	.140
It's a way to spend time instead of getting bored	46%	53%	38%	14.93	.000
It makes you feel grown up	42%	44%	41%	0.58	.445
It makes people see you as an adult	32%	34%	30%	1.09	.295

Table 2. Factor structure of Driving Functions scale

Item		Factor 1 LEISURE *Eigenvalue = 2.42* *% of Var. 25.80*	Factor 2 AUTONOMY *Eigenvalue = 1.32* *% of Var. 10.30*
V48	a way to spend time	**.69**	-.07
V54	a good opportunity to meet friends	**.68**	.00
V53	it's fun	**.66**	.03
V49	makes you feel grown up	**.35**	.15
V57	travel more independently	-.14	**.71**
V47	more independent from parents	.11	**.49**
V50	to get to more distant places	.19	**.39**
	Cronbach Alpha	.68	.60

Exploratory Factor Analysis on Driving Functions Scale

To assess the psychometric properties of Risky Driving Functions Scale, the items were subjected to a principal axis factor analysis with Promax rotation.

A first principal axis factor analysis with subsequent Promax rotation was conducted on 11 items of Driving Functions Scale. Promax rotation is a procedure normally used when factors are expected to correlate and not to be orthogonal (Gorsuch, 1983). The four items that presented a loading less than .30 on more than one factor were deleted.

A second principal axis factor analysis with Promax rotation was performed on the remaining 7 items. The scree test of eigenvalues (Cattell & Vogelmann, 1977) reveals two principal factors that accounted for 36.10% of the variance (see Table 2).

The results revealed a two-factor structure corresponding to two domains of Driving Functions named Leisure and Autonomy, that respectively accounted for 25.80% and 10.30% of the variance; the actual item loadings in the factors ranged from .35 to .71 (see Table 2).

The scales measuring the different aspects of Driving Function resulted in substantial correlations (r=.32). The Cronbach's Alpha coefficient for the Leisure scale was .68, while for the Autonomy scale was .60 (see Table 2).

Confirmatory Factor Analysis on Driving Functions Scale

To further investigate the relations among the two factors, we examined the hypothesis that Leisure and Autonomy Functions scales, although correlated, assess independent constructs. To this aim we performed confirmatory factor analyses (using the M-Plus Program - 4.1 Version; Muthén and Muthén, 2007). Three models were tested.

1) Unique factor model: in this model, factor loadings and error variances are estimated when the correlations between factors are fixed at 1. The model hypothesizes that the Driving Function scales are the expression of one latent factor.

Table 3[a]. Fit indices for confirmatory factor analyses

Models	χ^2	df	p-value	CFI	RMSEA	C. I.	SRMR
A. Unique factor	395.365	14	< .001	.41	.12	(.10 .14)	.11
B. Orthogonal model	100.511	14	< .001	.87	.10	(.08 .12)	.11
C. Oblique model	29.565	13	< .001	.97	.04	(.02 .07)	.04

[a] All the models yield a significant χ^2 at the $p < .001$ level; CFI = Comparative fit index; RMSEA = Root means square error of approximation; C.I. = Confidence Interval; SRMR = Standardized Root Mean Square Residual.

2) Orthogonal model: in this model, factor loadings and error variances are estimated when the correlations between factors are fixed at 0. The model hypothesizes that the Driving Function scales are independent.
3) Oblique model: In this model, factor loadings and error variances are estimated and the correlations between factors are estimated. The model hypothesizes that the Driving Function scales are interrelated but not the expression of one latent factor.

Models'goodness of fit was evaluated by means of the χ^2 statistic (Joreskog & Sorbom, 1984), the Comparative Fit Index (CFI; Bentler, 1990), and the Root mean square error of approximation (RMSEA; Bentler, 1995).

Table 3 shows the different fit indices of the tested models. As can be seen, all models obtained a significant χ^2, likely due to the large sample size. Furthermore, the oblique model presents fully adequate fit indexes and fit the data better than the unique factor model and the orthogonal model. These results indicate that the scales refer to two distinct but correlated driving functions, Leisure and Autonomy.

Drivers/Non-Drivers Invariance

We investigated if the latent structures of the Driving Functions Scale were replicated for both drivers and non-drivers. Confirmatory factor analyses (CFAs) were computed for each sample, and a multiple-group approach (MGCFA) was used to test measurement invariance for drivers and non-drivers (using Mplus 4.1; Muthén & Muthén, 2007). Two consecutively more restrictive analyses of invariance were run with the maximum likelihood method of estimation. We tested two nested models (Vandenberg, 2002): Configural invariance (i.e., the same pattern of fixed and free factor loadings was specified for each group) and metric invariance (i.e., the same factor loadings for items were specified for each group),

To test measurement invariance χ^2 difference tests were performed to compare nested models, adopting a cut-off of $p < .01$ given that obtaining a significant χ^2 becomes increasingly likely with large sample sizes (Kline, 1998).

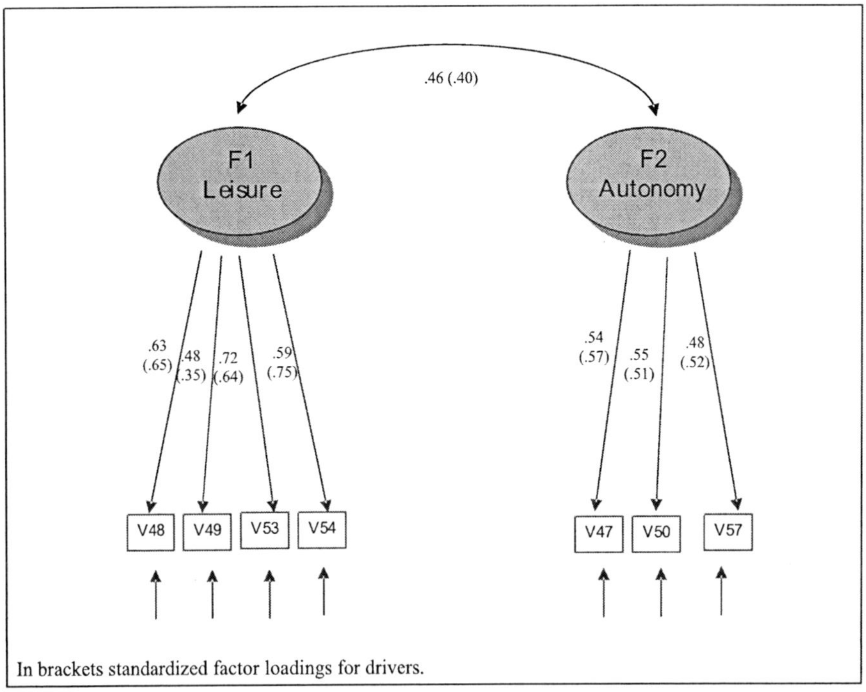

Figure 1. Path diagram of the Leisure and the Autonomy Scales

Table 4. Fit Indices for the Multi-Group (Drivers-Non-Drivers) CFAs

	χ^2	df	p-value	CFI	RMSEA	Conf. Int.	SRMR	χ^2_{diff}	df_{diff}	p-value
Configural invariance	67.959	26	<.001	.95	.06	(.04 - .08)	.06			
Metric Invariance	81.869	32	<.001	.95	.06	(.04 - .08)	.06	13.91	6	.03

The fit indices for the configural invariance models (Table 4) suggested that the same factor structure existed for drivers and non-drivers for both the Leisure and the Autonomy. The mean change in overall χ^2 between the configural invariance model and the metric invariance model was nonsignificant, χ^2_{diff} (6, N = 616) = 13,91, p = .03, suggesting that structures and factor loadings are the same in both drivers and non-drivers (see Table 4).

Risky Driving Functions

With regards to risky driving functions, results showed that the percentages of agreement with each item (sum of agree and strongly agree) varied between 38% and 5% (see Table 5).

The highest percentages of endorsement were observed for the items it's exciting (38%) and it's a way to relieve tension (33%). 23% agreed that risky driving is also a way to test driving skills. Therefore, the main functions of risky driving seem to concern the deliberate will to experience strong sensations and the thrill of risk.

Table 5. Risky driving functions: percentages of agreement (sum of *agree* and *strongly agree*)

Items	% Total	% Drivers	% Non-drivers	χ^2	p
It's exciting	38%	42%	33%	5.91	.015
It's a way to relieve my tension	33%	38%	25%	7.93	.005
It's a way to test how fast the car/motorcycle/moped goes	31%	31%	22%	13.14	.000
It's a way to have fun with friends	26%	31%	20%	10.01	.002
I like to take risks	25%	31%	19%	10.25	.001
It's a way to test my driving skills	23%	31%	16%	19.02	.000
It's a way to show others how fast the car/motorcycle/moped goes	21%	26%	16%	8.38	.004
It's a way to test if I can control the car/motorcycle/moped	17%	24%	10%	21.79	.061
It's a way to make a big impression on a girl/boy	15%	15%	15%	0.01	.933
It's a way to break the rules	14%	14%	14%	0.01	.940
It's a way to "show off" in front of my friends	13%	13%	12%	0.04	.835
I comply with the group	11%	11%	11%	0.07	.796
It's a way to show others I'm brave	9%	8%	9%	0.26	.610
I can't resist pressure from my friends to drive in a risky way	5%	6%	4%	0.57	.450

Reasons tied to peer relationships were less endorsed by adolescents (for example: I can't resist pressure from my friends, 5%, and to show off for friends 13%).

As for driving functions, some significant differences were found between drivers and non-drivers in the perception of risky driving functions. In particular, adolescent drivers assign to risky driving the functions of coping (a way to relieve tension), sensation-seeking (It's exciting; I like risk), and testing skills (It's a way to test my driving skills; it's a way to test - and show others - how fast the car/motorcycle/moped goes) more frequently than non-drivers.

Exploratory Factor Analysis on Risky Driving Functions Scale

A first principal axis factor analysis with subsequent Promax rotation was conducted on 14 items of Risky Driving Functions Scale. The two items that presented a loading less than .40 were deleted.

A second principal axis factor analysis with Promax rotation was performed on the remaining 12 items. The scree test of eigenvalues (Cattell & Vogelmann, 1977) revealed two principal factors that accounted for 48.70% of the variance (see Table 6). The two-factor structure corresponding to the two domains of Risky Driving Functions named Identity and

Relations, that respectively accounted for 42.50% and 6.20% of the variance; the actual item loadings in the factors ranged from .83 to .50.

The scales measuring the different aspects of Risky Driving Functions resulted in substantial correlations. In particular, Identity and Relations Functions showed a correlation of .69, statistically significant, attesting to a substantial overlap among the different aspects Risky Driving Function.

The Cronbach's Alpha coefficient for the Identity scale was .86 and for the Relations scale was .82.

Confirmatory Factor Analysis on Driving Functions Scale

To further investigate this overlap, the hypothesis that Identity and Relations Functions scales are correlated but assess independent constructs was examined by means of confirmatory factor analyses using the M-Plus Program 4.1 Version (Muthén and Muthén, 2007).

We tested three different models: unique factor, orthogonal model and oblique model.

Table 7 shows the different fit indices of the tested models. As can be seen, all models obtained a significant χ^2, likely due to the large sample size. Also, the oblique model presents fully adequate fit indexes and fits the data better than either the unique factor model or the orthogonal model. These results indicate that the scales are interrelated but they are not the expression of a single latent variable.

Table 6. Factor structure of Risky Driving Functions scale

Item		Factor 1 IDENTITY *Eigenvalue = 5.60* *% of Var. 42.50*	Factor 2 RELATIONS *Eigenvalue = 1.25* *% of Var.6.20*
V61	It's exciting	**0.75**	0.03
V71	like to take risks	**0.74**	-0.06
V63	to relieve my tension	**0.70**	-0.12
V58	to test how fast the car/motorcycle/moped goes	**0.66**	0.06
V69	to test my driving skills	**0.64**	0.13
V64	to control car/motorcycle/moped	**0.60**	0.12
V70	to break the rules	**0.50**	0.17
V67	to show others I'm brave	-0.10	**0.83**
V60	to "show off" in front of my friends	0.06	**0.72**
V66	to comply with the group	-0.03	**0.66**
V59	It can't resist pressure from my friends	0.10	**0.61**
V65	to make a big impression on a girl/boy	0.11	**0.58**
	Cronbach Alpha	.86	.82

Table 7 [a]. Fit indices for confirmatory factor analyses

Models	χ^2	*df*	p-value	CFI	RMSEA	C.I.	SRMR
A. Unique factor	493.452	53	< .001	.89	.11	(.11 .12)	.06
B. Orthogonal model	633.154	53	< .001	.86	.13	(.12 .13)	.28
C. Oblique model	207.774	52	< .001	.96	.06	(.05 .07)	.04

[a] All the models yield a significant χ^2 at the $p < .001$ level; CFI = Comparative fit index; RMSEA = Root means square error of approximation; C.I. = Confidence Interval; SRMR = Standardized Root Mean Square Residual.

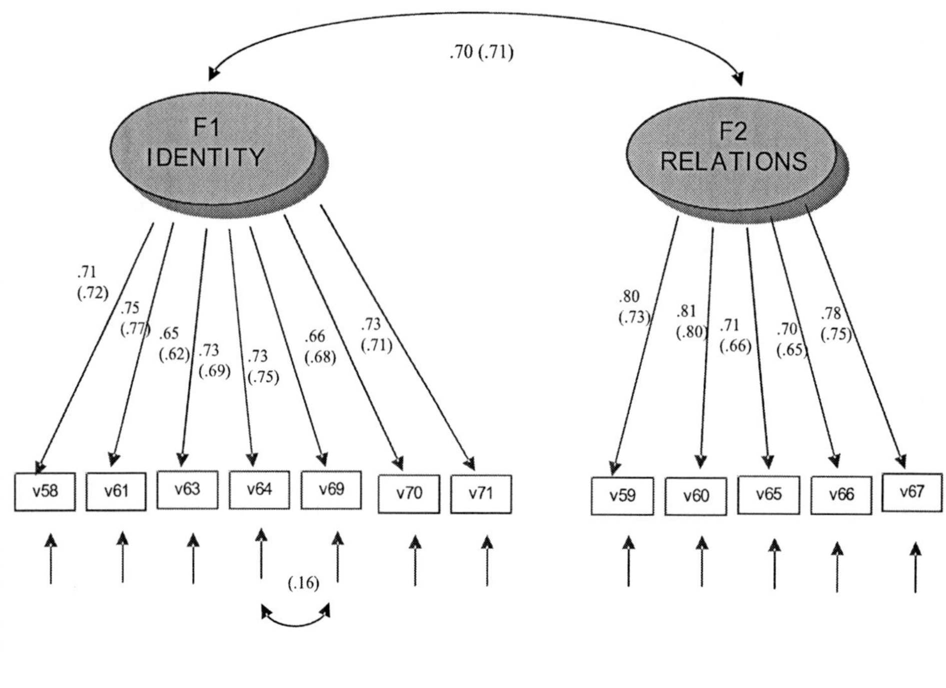

In brackets standardized factor loadings for drivers. The intercorrelation of the error terms between items v64 and v69 was estimated.

Figure 2. Path diagram of the Identity and the Relations Scales

Drivers/Non-Drivers Invariance

We investigated if the latent structures of the Risky Driving Functions Scale were replicated for both drivers and non-drivers.

The fit indices for the configural invariance models suggested that the same factor structure existed for drivers and non-drivers for both the Identity and the Relations. The mean change in overall χ^2 between the configural invariance model and the metric invariance model was nonsignificant, χ^2_{diff} (11, N = 616) = 14.41, p = .21, suggesting that factorial structure and factor loadings of the two scales are the same in drivers and non-drivers (see Table 8).

Driving and Risky Driving Functions and Implication in Risky Driving

With regard to our last aim, preliminary correlation analyses were carried out (see Table 9). With regard to driving functions, correlation coefficients showed a significant but weak relationship between leisure functions and involvement in risky driving (r=.17). No significant relation was found between autonomy functions and involvement in risky driving.

In terms of risky driving functions, both social functions and identity-related functions were significantly associated with greater involvement in risky driving, though the correlation between identity-related functions and risky driving was stronger (r=.45) than the one between social functions and risky driving (r=.27). Thus, the two most strongly correlated factors of risky driving functions (social relationships and identity) turned out to be associated with involvement in risky driving to different degrees.

Hierarchical multiple regression analysis (see Table 10) was carried out to evaluate the joint effects of driving and risky driving functions on involvement in risky driving and the specific contribution of each different factor of these functions.

In the hierarchical model, the impact of different factors of driving and risky driving functions on risky driving was evaluated after partialling out the effect of weekly mileage and type of vehicle driven (first step) and the effects of gender, age, and type of school (second step). In the last step, the two factors of driving functions (leisure and autonomy) and the two factors of risky driving functions (social and identity-related functions) were entered as predictors (see Table 10).

Table 8. Fit Indices for the Multi-Group (Drivers-Non-Drivers) CFAs

	χ^2	df	p-value	CFI	RMSEA	Conf. Int.	SRMR	χ^2_{diff}	df_{diff}	p-value
Configural invariance	296.18	104	<.001	.95	.07	(.07 - .08)	.04			
Metric invariance	310.59	115	<.001	.95	.06	(.05 - .08)	.05	14.41	11	.21

Table 9. Correlations

	Risky driving	Leisure functions	Automomy functions	Identity functions	Social functions
Risky driving	1	.17*	.12	.45**	.27**
Leisure functions	.17*	1	.29**	.44**	.40**
Automomy functions	.12	.29**	1	.10*	-.03
Identity functions	.45**	.44**	.10*	1	.63**
Social functions	.27**	.40**	-.03	.63**	1

* p <.05 ***p<.001

Table 10. Hierarchical multiple regression analysis

Step	Predictors		*β*	*Final β*	R^2	F	p	ΔR^2
1	Mileage		.37***	.22**	.25	16.74	.000	.25***
	Driving car		-.02	-.05				
	Driving motorcycle		.06	.05				
	Driving moped		.25**	.14*				
2	Gender		.21**	.12	.34	14.74	.000	.09***
	Age		.27**	.22**				
	Type of school		-.16*	-.12				
3	Driving functions:	Leisure		-.12	.41	12.72	.000	.07***
		Autonomy		.12				
	Risky driving functions:	Identity		.34***				
		Social		.01				

* p <.05 ** p < .01 ***p<.001

The final model accounted for 41% of the variance in risky driving involvement. The variables that independently and significantly predicted risky driving were weekly mileage, driving a moped, age, and identity-related functions of risky driving; this last variable was the best predictor. The effect of mileage and type of vehicle driven accounted for 25% of the variance; more specifically, greater involvement in risky driving was related to increase weekly mileage and to the use of a moped. Introducing gender, age, and type of school, a significant increment in explained variance (ΔR^2=.09; p<.001) was observed. Boys and students of vocational and technical schools in particular appeared to be more involved in risky driving and increasing involvement, in relation to age, was also observed. The addition of predictors related to different factors of driving and risky driving functions resulted in a further significant improvement in explained variance (ΔR^2=.07, p<.001), but only identity-related functions of risky driving revealed a significant relationship with involvement in risky driving. A positive beta value (β=.34) indicates greater involvement in risky driving especially for adolescents who assign identity-related functions to risky driving. Simultaneously introducing the two factors of risky driving functions in a regression model, social functions did not appear to be related to greater involvement in risky driving. Furthermore, regardless of risky driving functions, functions of driving (leisure and autonomy) were also found to be unrelated to risky driving.

Conclusion

The present study investigated risky driving with reference to the problem behavior approach, focusing on the concept of behavior functionality. According to Jessor (1987, 1998), risky driving, like other health risk behavior, is considered functional towards the attainment of relevant goals in the adolescents' life experience. The present study also investigated driving a motor vehicle (not necessarily in a risky way) as a functional behavior for adolescents. In fact, previous research has underlined that the use of cars and motor vehicles implies some benefits for adolescents, who can travel more easily, get more quickly to more distant places, gain greater independence, acquire status within their group of friends, and display their "adulthood" (Møller, 2004; Bonino et al. 2005).

The results of the present study stressed that driving and risky driving offer different advantages from the adolescents' point of view. According to previous research (Jessor et al., 1991; Møller, 2004; Bonino et al. 2005; Bina et al. 2008; Møller & Gregersen, 2008), these behaviors appeared to be functional in coping with adolescents' developmental tasks.

First considering driving, the advantages of driving a motor vehicle were agreed upon by the large majority of adolescents. Confirmatory factor analysis showed that these benefits referred to two different domains: personal autonomy on one hand, and chances to meet friends and have fun during leisure time, on the other hand. This evidence suggests that driving could have an important role in specific developmental tasks, such as the development of personal independence in adolescence and the enlargement of adolescents' social world and relationships, mainly with peers. Driving may also fulfill functions of adulthood, though less than half of the participants agreed with the statement that driving a motor vehicle makes them feel grown up. The results of factor analysis stressed that this feeling of adulthood is related more to the opportunity to have fun and meet friends freely than to the achievement of independence from parents and the ability to travel independently (in fact, this item loaded on the leisure factor).

Whereas the advantages of driving were shared by the majority of boys and girls, a smaller number of participants assigned positive functions to risky driving. However, more than a quarter of adolescent affirmed that risky driving could be a means to achieve specific goals, like experiencing risk and strong sensations, testing driving skills, or releasing tension. Furthermore, risky driving appeared to fulfill some functions related to reputation, status, visibility, acceptance, and identification in the context of the peer group. Factor analysis confirmed two factors of risky driving functions, in both subgroups of drivers and non-drivers. According to the original hypothesis, the first factor refers to functions concerning peer relationships, whereas the second factor refers to functions related to identity definition. Thus, on one hand, adolescents may fail to respect the traffic code and take risks while driving in order to be accepted by the peer group. On the other hand, driving fulfills functions of self-experimentation, related to the process of constructing an adult identity. Through risky driving, adolescents experience new and strong sensations, face risks, perceive control (although illusory) of a situation, test their skills and their limits, and transgress against social norms and laws. Our results stressed a strong correlation between the two different factors of risky driving functions. This finding confirmed our expectations that the process of constructing an adult and stable identity does not, in fact, occur within a "social vacuum", but rather in the context of social relationships. Thus, identity related functions were strongly interrelated with the social functions of risky driving. For instance, improving their reputation, gaining visibility, and demonstrating toughness and bravery by taking risks while driving allows adolescents to improve their status in the peer group, while at the same time, it affirms and strengthens their identity, also through the support of peers.

These results have, however, some limitations. First of all, the study did not consider some instrumental functions of driving, such as time management, for instance. Previous research has stressed that these types of reasons were the main justification for traffic violations in a sample of adolescent girls (Shapiro et al., 1998). Our study focused only on risky driving functions related to developmental tasks and the problems that adolescents face in this period of their life. Further research should also consider instrumental functions of driving as well as development-related functions. The main limitation of the study is tied to the fact that we investigated driving and risky driving functions as perceived by adolescents. This way of investigating behavior functions emphasizes the more conscious reasons for behavior. To better understand risky driving in adolescence it is also important to investigate the less conscious reasons for this type of behavior. An indirect investigation of behavior functions (Jessor at al., 1991; Bonino et al. 2005) might also be able to highlight reasons that are not fully mindful, taking into consideration the relationships between implication in risk behavior and individual and contextual characteristics. For this reason, future research should combine direct and indirect methods of investigating driving and risky driving functions to reach a more complete understanding of these constructs.

Despite these limitations, we recognize the importance of investigating risk behavior functions by asking adolescents their motivations and opinions directly, since, from a constructivistic point of view, it has been repeatedly demonstrated that the individual's perception of reality is an important source of information in understanding individual action (Magnusson & Stattin, 1997; Juang & Silbereisen, 2001). Moreover, the present study highlights the importance of studying behavior functions perceived by adolescents directly also in terms of methodology. In fact, obtaining data on the functions that adolescents assign to driving and risky driving allows us to analyze the relations between these functions and involvement in risky driving itself.

Investigating these relations, the present study found some interesting results. With regard to the role of driving functions in risky driving, the leisure function was found to be moderately related to involvement in risky driving. Indeed, adolescents who more frequently assigned leisure functions to driving were more implicated in traffic violations and risk-taking while driving. However, when analyzed in a multiple regression model including driving-related variables (mileage and vehicles driven), socio-demographic variables and risky driving functions, this relation became non-significant. Further analysis is needed to evaluate if these variables, and in particular risky driving functions, could have mediation effects on the relation between the leisure function of driving and involvement in risky driving. Moreover, this result suggests that the relations between driving, risky driving, and leisure time activity should be further investigated, for instance by taking into consideration differences in adolescents' leisure activities. In fact, previous research has indicated that different ways of spending time with friends are likely to be associated with different adolescent driving behavior. Different studies have stressed that adolescent lifestyle, also in terms of leisure time activities, is strongly related to risky driving behavior (Chlioutakitis, Koukouli, Lajunen & Tzamalouka, 2005; Bina, Graziano & Bonino, 2006; Møller & Gregersen, 2008) and accident risk (Berg, 1994; Gregersen & Berg, 1994; Chlioutakitis, Darviri & Demakakos, 1999).

With reference to risky driving functions, both social and identity-related functions, were correlated with greater involvement in risky driving. However, when the two scales were entered into a multiple regression model simultaneously, only the identity-related functions scale showed a significant relation with risky driving. Thus, adolescents who take risks while driving as a way to test their abilities, to experience strong sensations, and to perceive control over their actions, are the ones who are most involved in risky driving. These adolescents are probably experiencing greater difficulty in the construction of their adult identity and they use risky driving as a means to cope with this fundamental developmental task.

The relations between different factors of risky driving functions and implication in this behavior should also be investigated further. A relevant variable to consider might be whether deliberate traffic violations and risk-taking while driving occur particularly when driving with friends or alone or in both situations. It is likely that young people who take risks while driving with reference to social reasons behave in this way when they are with peers, but not necessarily when they are alone. This may be one of the possible reasons for the weaker relationship between social function and risky driving, though further analysis is needed to better understand this topic.

The results of the present study could not be fully compared to earlier research on the functions, benefits, and reasons for driving and risky driving. Previous studies have considered a different group of motives and justifications for risky driving (Berg, 1994; Gregersen & Berg, 1994; Shapiro et al. 1998) or have investigated the relations between taking risks while driving and driving functions without differentiating the specific functions of risky driving (Møller & Gregersen, 2008). However, these studies do support the evidence that some functions, though other terminology was used (see introduction), related to identity and social relationships with peers were related to greater involvement in risky driving and accident risk.

The results concerning the relationships between functions and risky driving involvement also have some limitations. First of all, the measures were self-reported. While the limitations related to investigating the perceived functions of driving and risky driving were outlined previously, with regard to risky driving measures, previous studies have demonstrated the reliability of the self-report of driving behavior (Hatakka, Keskinen, Katila & Laapotti, 1997; Lajunen & Summala, 2003). A second limitation is linked to the cross-sectional design of our research. We recognize the need to develop a longitudinal research design in order to better investigate the potential risk or protective role of driving and risky driving functions and the stability or variability of such functions for the same person during adolescence. Moreover, in this study we did not take into consideration possible gender and age-related differences in the perceived functions of driving and risky driving. These differences should be considered in further studies. Finally, this study was conducted in northwestern Italy and caution must be used when generalizing these results to adolescents with different demographic characteristics or coming from different countries. While driving may assume different cultural meanings in different countries, some characteristics of driving behavior in Italian adolescents are likely to be similar to those of adolescents from other European countries. For instance, adolescents are over-represented in traffic accidents in Italy as well as in other European countries (ERSO, 2008). As shown by international studies, this evidence suggests that driving behavior and risky driving in adolescence are related to developmental factors and problems. Moreover, the above-mentioned research, which investigated motives, reasons, justifications, and functions of driving and risky driving behavior in young people found similar reasons for such behavior (Shapiro et al., 1998; Møller, 2004; Møller & Gregersen, 2008). Therefore, our results appear to confirm the results of earlier research. Further studies are also needed to understand the specific role of cultural factors with respect to the functions that Italian adolescents assign to driving and risky driving.

Despite these limitations, the results of our study provide some important suggestions for risky driving prevention programs targeting adolescent drivers. First of all, our results support the importance of implementing intervention programs which take into consideration the meanings and functions of driving and risky driving in adolescence, and the strong relationships between this behavior and adolescent developmental tasks (Bonino et al. 2005). The relationships between social functions and risky driving lend support for peer group based interventions (Ferrer-Wreder, Stattin, Cass Lorente, Tubman & Adamson, 2004). Moreover, the evidence that adolescents who assign identity-related functions to risky driving are more involved in this behavior, suggests that these boys and girls should be supported in the process of constructing a stable adult identity. Interventions should be focused on the promotion of specific skills (Ferrer-Wreder et al., 2004) to cope effectively with developmental problems and tasks, as suggested by other studies (Møller & Gregersen, 2008). Furthermore, interventions should allow adolescents to find alternative strategies and more safe behaviors to reach the same significant goals achieved through risky driving (Bonino & Cattelino, 2008).

Further suggestions are tied to the evidence that risky driving is related to the type of vehicle driven. Specifically, greater involvement was found to be related to driving a moped. While risky driving appears to be more frequent in late adolescence, adolescents who had driven a moped (in Italy this is typical of minors) were more involved in risky driving. Also in light of other studies which have highlight the risks related to moped driving (Kopjar, 1999; Michel et al., 2002), this matter certainly deserves further investigation. Finally, the results of the present research support the need to implement differentiated prevention and intervention programs, targeting different groups of adolescents and taking into consideration age, motivations, and type of motor vehicle driven.

Acknowledgments

The authors are grateful to Prof. Silvia Bonino for her precious supervision of the entire research project, to Valentina Aprile for her contribution to preliminary data analysis, and to Lindsay Mc Donald for her English revision of the manuscript.

References

Asbridge, M., Poulin, C. & Donato A. (2005). Motor vehicle collision risk and driving under the influence of cannabis: evidence from adolescents in Atlantic Canada. *Accident Analysis and Prevention, 37*, 1024- 1034.

Bentler, P. M. (1990). Comparative fit indexes in structural models. *Psychological Bulletin, 107*, 238–246.

Bentler, P. M. (1995). *EQS: Structural equations program manual, Version 5.0.* Los Angeles: BMDP Statistical Software.

Berg, H.Y. (1994). *Lifestyle, traffic and young drivers. An interview study.* VTI-report, 389 A, Sweden.

Bina, M., Graziano, F. & Bonino, S. (2006). Risky driving and lifestyle in adolescence. *Accident analysis and Prevention, 38*, 472-481.

Bina, M., Graziano, F., Borca, G. & Calandri, E. (2008). The psychological functions of risky driving in adolescence: the adolescents' point of view. In A. De Smet (Ed.), *Transportation Accident Analysis and Prevention* (pp. 1-24). New York: Nova Science Publishers Inc.

Bonino, S. & Cattelino, E. (Eds.) (2008). *La prevenzione in adolescenza [Prevention in adolescence].* Trento (Italy): Erickson.

Bonino, S., Cattelino, E. & Ciairano, S. (2005). *Adolescents and risk. Behaviors, functions and protective factors.* New York: Springer Verlag.

Cattell, R. B. & Vogelmann, S. (1977). A comprehensive trial for the scree and KG criteria for determining the number of factors. *Multivariate Behaviour Research, 12*, 289–325.

Chlioutakitis, J. E., Darviri, C. & Demakakos, P. T. (1999). The impact of young drivers' lifestyle on their road traffic accident risk in greater Athens area. *Accident Analysis and Prevention, 31*, 771-780.

Chlioutakitis, J. E., Koukouli, S., Lajunen, T. & Tzamalouka, G. (2005). Lifestyle traits as predictors of driving behavior in urban areas of Greece. *Transportation Research, F8*, 413-428.

Ciairano, S. (2004). *Risk behaviour in adolescence: drug-use and sexual activity in Italy and the Netherlands.* Groningen, The Netherlands: Stichting Kinderstudies Publisher.

Cooper, M. L. (1994). Motivations for alcohol use among adolescents: development and validation of a four-factor model. *Psychological Assessment, 6, 2*, 117-128.

Cooper, M. L., Shapiro, C. M. & Powers, A. M. (1998). Motivations for sex and risky sexual behavior among adolescents and young adults: a functional perspective. *Journal of Personality and Social Psychology, 75*, 1528-1558.

Costa, F. M., Jessor, R. & Turbin, M. S. (2007). College student involvement in cigarette smoking: The role of psychosocial and behavioral protection and risk. *Nicotine and Tobacco Research,* 9(2), 213-224.

Engels, R. (1998). *Forbidden fruits. Social dynamics in smoking and drinking behavior of adolescents.* Maastricht: Universitaire Pers Maastricht.

Engels, R. & ter Bogt, T. (2001). Influences of risk behaviors on the quality of peer relations in adolescence. *Journal of Youth and Adolescence, 30*, 675-695.

ERSO (European Road Safety Observatory) (2008). *Annual Statistical Report.* http://www.erso.eu

Ferrer-Wreder, L., Stattin, H., Cass Lorente, C., Tubman, J. G. & Adamson, L. (2004). *Successful prevention and youth development program. Across borders.* New York: Kluwer Academic/Plenum Publishers.

Gebhardt, W. A., Kuyper, L. & Greunswen, G. (2003). Need for intimacy in relationships and motives for sex as determinants of adolescents condom use. *Journal of Adolescent Health, 33*, 154-164.

Giannotta, F., Ciairano, S., Bonino, S., Morero, D. (2004). Le funzioni dell'uso di sostanze psicoattive e dei rapporti sessuali: il punto di vista degli adolescent [The functions of drug use and sexual intercourses: the adolescents' point of view]. *Psicologia della Salute, 3*, 23-37.

Giannotta, F., Ciairano, S., Spruijt, R., Spruijt-Metz, D. (in press). Meanings of sexual intercourse for Italian adolescents. *Journal of Adolescence* (2008), doi: 10.1016/j. adolescence. 2008.05.009.

Gorsuch, R. L. (1983). *Factor Analysis.* (2nd Edn. ed.). New York: Plenum Press.

Gregersen, N. P. & Berg, H.Y. (1994). Lifestyle and accidents among young drivers. *Accident Analysis and Prevention, 26*, 297-303.

Hatakka, M., Keskinen, E., Gregersen, N. P., Glad, A. & Hernetkoski, K. (2002). From control of the vehicle to personal self-control; broadening the perspectives to driver education. *Transportation Research Part F, 5*, 201–15.

Hatakka, M., Keskinen, E., Katila, A. & Laapotti, S. (1997). Self-reported driving habits are valid predictors of violations and accidents. In T. Rothengatter & E. Carbonell Vaya (Eds.), *Traffic and transport psychology. Theory and application* (pp. 295-303). Pergamon.

Hsia, F., N., Spruijt-Metz, D. (2003). The meanings of smoking among Chinese American and Taiwanese American college students. *Nicotine and Tobacco Research, 5*, 1-14.

ISTAT (National Institute of Statistics)/ACI (Automobile Club Italy) (2007). *Incidenti stradali. Anno 2005 [Traffic accidents statistics, 2005]*. Rome: ISTAT.

Jessor, R. (1987). Risky driving and adolescent problem behavior: an extension of problem-behavior theory. *Alcohol, Drugs and Driving, 3*, 1–11.

Jessor, R. (Ed.) (1998). *New Perspectives on Adolescent Risk Behaviour*. Cambridge: Cambridge University Press.

Jessor, R., Donovan, J. E. & Costa, F. M. (1991). *Beyond Adolescence. Problem Behavior and Young Adult Development*. New York: Cambridge University Press.

Jessor R., Jessor S. L. (1977). *Problem behavior and psycosocial development: A longitudinal study of youth*. New York: Academic Press.

Jessor, R., Turbin, M. S. & Costa, F. M. (1997). Predicting developmental change in risky driving: the transition to young adulthood. *Applied Developmental Science, 1 (1)*, 4-16.

Joreskog, K. G. & Sorbom, D. (1984). *LISREL VI: Analysis of linear structural relationships by method of maximum likelihood*. Chicago: National Educational Resources.

Juang, L. P. & Silbereisen, R. K. (2001). Family transitions for young adult women in the context of a changed Germany: timing, sequence and duration. *American Behavioral Scientist, 44*, 1899-1917.

Kline, R. B. (1998). *Principles and practices of structural equation modeling*. New York: Guilford.

Kloep, M., Guney, N., Cok, F. & Simsek, O.F. (in press). Motives for risk-taking in adolescence: a cross-cultural study. *Journal of Adolescence* (2008), doi: 10.1016/j.adolescence. 2007.10.010.

Kopjar, B. (1999). Moped injuries among adolescents: a significant forgotten problem? *Accident Analysis and Prevention, 31*, 473-478.

Laapotti, S., Keskinen, E., Hatakka, M. & Katila, A. (2001). Novice drivers' accidents and violations – a failure on higher or lower hierarchical levels of driving behaviour. *Accident Analysis and Prevention, 33*, 759-769.

Lajunen, T. & Summala, H. (2003). Can we trust self-reports of driving? Effect of impression management on driving behaviour questionnaire responses. *Transportation Research, F-6*, 97-101.

Magnusson, D. & Stattin H. (1997). Person-context interaction theories. In R. Lerner & W. Damon (Eds.), *Theoretical models of human development, vol. 1, Handbook of child psychology*. New York: Wiley.

Michel, G., Purper-Ouakil, D. & Mouren-Siméoni, M. C. (2002). Risk taking in teenagers. Dangerous moped driving. *Neuropsychiatrie de l'enfance et de l'adolescence, 50*, 583–589

Møller, M. (2004). An explorative study of the relationship between lifestyle and driving behaviour among young drivers. *Accident Analysis and Prevention, 36*, 1081-1088.

Møller, M. & Gregersen, N. P. (2008). Psychosocial function of driving as predictor of risk-raking behaviour. *Accident Analysis and Prevention, 40*, 209-215.

Muthen, L. K. & Muthen, B. O. (2007). *Mplus 4.1 User's Guide*. Los Angeles: Muthen & Muthen.

Newcomb, M. D., Chou, C. P., Bentler, P. M., Huba, G. J. (1988). Cognitive motivations for drug use among adolescents: longitudinal tests of gender differences and predictors of change in drug use. *Journal of Counseling Psychology, 35, 4*, 426-438.

Robinson, M. L., Holmbeck , G. N. & Paikoff, R. (2007). Self-esteem enhancing reasons for having sex and the sexual behaviors of African American adolescents. *Journal of Youth and Adolescence, 36*, 453-464.

Shapiro, R., Siegel, A. W., Scovill, L. C. & Hays, J. (1998). Risk-taking patterns of female adolescents: what they do and why. *Journal of Adolescence, 21*, 143-159.

Shope, J. T., Raghunathan, T. E. & Patil, S. M. (2003). Examining trajectories of adolescent risk factors as predictors of subsequent high-risk driving behaviour. *Journal of Adolescent Health, 32*, 214-224.

Silbereisen, R. K., Eyferth, K., Rudinger, G. (1986). *Development as action in context.* Berlin: Springer-Verlag.

Silbereisen, R. K., Noack, P. (1988). On the constructive role of problem behaviour in adolescence. In: Bolger, A., Caspi, N., Downey, G., Moorehose, M. (Eds.), *Person in context: Development process* (pp. 152-180). Cambridge, England: Cambridge University Press.

Spruijt-Metz, D. (1999). *Adolescence, affect and health.* Hove, UK: Psychology Press.

Spruijt-Metz, D., Gallagher, P. E., Unger, J. B., Anderson-Johnson, C. (2004). Meanings of smoking and adolescent smoking across ethnicities. *Journal of Adolescent Health, 35*, 197-205.

Spruijt-Metz, D., Gallagher, P. E., Unger, J. B., Johnson, C. A. (2005). Unique contributions of meanings of smoking and outcome expectancies to understanding smoking initiation in middle school. *Annals of Behavioral Medicine, 30 (2)*, 104-111.

Tilleczek, K. C. & Hine, D. W. (2006). The meaning of smoking as health and social risk in adolescence. *Journal of Adolescence, 29*, 273-287.

Vandenberg, R. J. (2002). Toward a Further Understanding of and Improvement in Measurement Invariance Methods and Procedures. *Organizational Research Methods,* 5, 139-158.

In: Accidents: Causes, Analysis and Prevention
Editors: Henri Bédard and Geraud Delashmit
ISBN: 978-1-60741-712-5

Chapter 3

THE ROLE OF PARENTS AND FRIENDS IN ADOLESCENT RISKY DRIVING: HOW THEY DRIVE AND WHAT THEY THINK ABOUT RISKY DRIVING

Federica Graziano**, Manuela Bina and Tatiana Begotti
Department of Psychology, University of Turin, Turin, Italy

ABSTRACT

Developmental psychology literature has stressed the role of parents' and friends' behavior models and attitudes in adolescents' involvement in risk behavior (Jessor et al., 1991; Bonino et al., 2005). With reference to risky driving (traffic code violations and driving in a risky way) different studies have investigated the role of parents' and friends' driving style as well as their attitudes about adolescent involvement in risky driving (Hartos et al., 2000; Taubman-Ben-Ari et al., 2005; Wilson et al., 2006). However, previous studies have only examined these predictors separately. The present study investigates the combination of these variables in the prediction of self-reported risky driving, by taking into account the differences between the maternal and paternal role and the roles of friends and the best friend. Evaluations of parents' and friends' risky driving models and their disapproval of risky driving are based on adolescents' perceptions.

The main aim of the study was to investigate the relationship between parents' risky driving models and risky driving disapproval (looking separately at mothers and fathers), friends' risky driving model and risky driving disapproval (distinguishing between friends and best friend) and adolescents' involvement in risky driving. Gender differences were also investigated. We expected that adolescents' greater involvement in risky driving was related to parents' and friends' risky driving model, while lower involvement was related to parents' and friends' disapproval of risky driving.

Data were collected through an anonymous self-report questionnaire. The study involved 341 adolescent drivers (car, motorbike, moped) (50% boys), ages 15 to 19, attending different high schools in Piedmont (northwestern Italy). Risky driving was evaluated using the scale developed by Bonino et al. (2005), while items related to parents' and friends' risky driving model and risky driving disapproval were formulated specifically for this study. Correlations and hierarchical multiple regressions were carried

* Corresponding author: fgrazian@psych.unito.it

out, respondent age was controlled for, and gender and average weekly miles driven were considered in order to investigate relationships between variables.

Main results stressed that: 1) in the total sample significant predictors of risky driving were gender, paternal, friends' and best friend's risky driving models, and friends' disapproval of risky driving; 2) for boys the best predictor was friends' disapproval of risky driving followed by friends' and paternal risky driving model; 3) for girls significant predictors were age and best friend's risky driving model.

Implications for risky driving prevention in adolescence are discussed.

INTRODUCTION

Driving a car or riding a motorcycle generally produces significant changes in an adolescent's lifestyle (Williams, 1998); it allows for greater freedom of movement, independence from parents, and opportunities to meet friends. Moreover, driving a motor vehicle fulfills many positive psychosocial functions related to adolescent developmental tasks, such as self-affirmation and acquisition of physical and psychological autonomy from parents (Bonino, Cattelino & Ciairano, 2005), as well as visibility among friends, status, control, and mobility (Møller, 2004; Møller and Gregersen, 2008). Nonetheless, driving may be dangerous for the health and safety of adolescents especially when they take risks while driving and fail to respect the traffic code (Jessor, 1987, 1998; Shope, Raghunathan & Patil, 2003; Asbridge, Poulin & Donato, 2005; Bonino et al., 2005). Motor vehicle accidents are the main cause of death among 15- to 34-year-olds in Italy and in other European countries (ISTAT/ACI, 2007; ERSO, 2008), and among young people risky driving style is, in fact, the principal cause of motor vehicle accidents. Risky driving style refers to the violation of traffic codes, particularly the failure to respect speed limits, stop signals, and safety distances. Young people are more likely to overtake recklessly and to fail to use safety devices, such as crash helmets and seat belts. They are also more likely to drive under the influence of alcohol and other psychoactive substances (Jonah, 1990; Hartos, Eitel, Haynie & Simons-Morton, 2002; Williams, 2003; Bonino et al., 2005). An adolescent's probability of having a road accident is linked not so much to mileage, as to the violation of traffic codes; in other words the risk is greater for individuals who drive in a risky manner than for individuals who drive a great deal (Bonino et al., 2005).

The problem of risky driving in adolescence has been examined and interpreted by the psychological literature from different points of view. Many studies have stressed the role of individual variables, particularly personality traits such as impulsiveness and aggressiveness (Beirness, 1993; Deffenbacher, Lynch, Oetting & Yingling, 2001), lack of experience in driving (Finken, Jacobs & Laguna, 1998; Williams, 1998), and risk proneness and the need for sensation-seeking (Jonah, 1986; Yu & Williford, 1993; Arnett, 1996; Arnett, Offer & Fine, 1997; Vavrik, 1997). Developmental psychology studies have stressed that risky driving is not an isolated behavior and that it should be investigated in relation to numerous aspects of the adolescent lifestyle (Jessor, Turbin & Costa, 1997). In other words, risky driving should be considered with reference to a complex system comprised of individual variables (such as values, opinions, attitudes, and motivations) (Laapotti, Keskinen, Hatakka & Katila, 2001; Hatakka, Keskinen, Gregersen, Glad & Hernetkoski, 2002), as well as behaviors and experiences related to adolescent life contexts, in particular family, peers, and school (Jessor,

1987; Jessor et al., 1997). In this study the specific role of parents and friends in relation to adolescents' risky driving will be examined.

Interest in parental influence on adolescent driving behavior has increased remarkably over the last years. Many studies have found statistically significant correlations between adolescents' and their parents' driving style (Taubman-Ben-Ari, Mikulincer & Gillath, 2005; Wilson, Meckle, Wiggins & Cooper, 2006) and some researchers have theorized an "intergenerational transmission" of driving style (Bianchi & Summala, 2004). While many studies have confirmed the relationship between parental and adolescent driving styles, the mechanisms of parental influence on adolescents' driving style is still not fully understood (Ferguson, Williams, Chapline, Reinfurt & De Leonardis, 2001). On one hand, some research has hypothesized that driving style is passed on from parents to adolescents through a genetic predisposition, which takes the form of temperament characteristics such as aggressiveness, impulsiveness, or sensation-seeking (Zuckerman, 1993; Koopmans, Boomsma, Heath & Vandoornen, 1995). Social learning theories, on the other hand, have suggested that the transmission of driving style should be seen as a learning process that centers around parents' behaviors and attitudes, which can have a critical impact on adolescents (Kandel & Andrews, 1987). Based on this point of view, many studies have claimed that parents are significant models for driving behavior. The influence of the parental driving model starts very early, when children first begin to observe their parents while driving, and continues throughout childhood and adolescence, particularly as they learn to drive under their parents' supervision (Bianchi & Summala, 2004). Some studies have focused on the specific influence of paternal and maternal driving models on adolescent driving, but results remain controversial. In general, adolescents are more likely to adopt a risky driving style if both parents are risky drivers, while a parent who drives safely can balance the influence of the other parent (Wilson et al., 2006). Some research has pointed to a stronger influence from the same sex parent (Harris & Morgan, 1991), while others have attained more complex results. For example, Taubman-Ben-Ari et al. (2005) argued that the paternal driving model has a significant role on boys' and girls' driving style, while the maternal driving model has a greater influence on girls' driving behavior. The importance of the paternal model has also been found in studies carried out in Italy. Bonino et al. (2005) found that adolescents are more likely to use seat belts if their fathers also do so. This aspect is particularly relevant because the use of safety devices (such as seat belts and crash helmets) is closely related to a cautious driving style. On the contrary, Wilson et al. (2006) suggested that boys and girls are more likely to violate speed limits if their mothers, but not their fathers, do so. In other words, mothers who violate the speed limit seem to be a greater risk factor than fathers who commit the same violation. The fact that driving fast is generally more widespread and accepted among males than females could explain the greater influence exerted, in this case, by the mother's behavior.

Parental attitudes towards driving practices can also have an influence on adolescent driving style. In the literature this aspect has traditionally received less attention or been neglected altogether, though it has been considered in some recent studies. These studies have found that parents can have a protective influence on adolescent involvement in risky driving not only by providing a cautious driving model, but also through explicit disapproval of risky driving (Bonino et al., 2005). In general, a lack of parental disapproval of a risk behavior may be interpreted by adolescents as implicit approval of this behavior. Moreover, parental attitudes are linked to the values, social norms, and messages they transmit related to risky driving (Taubman-Ben-Ari et al., 2005). For example, traffic code offences constitute a violation of institutional norms; thus, violating traffic codes could reflect a scant consideration of or refusal to abide by laws and these attitudes towards norms may be transmitted to adolescents by their parents (Wilson et al., 2006). With reference to norms, many studies have stressed the role of parental supervision and control in adolescent driving (Ferguson et al., 2001; Hartos et al., 2002; Bianchi & Summala, 2004; Bonino et al., 2005; Nygaard & Grube, 2005; Williams, Leaf, Simons-Morton & Hartos, 2006). The protective factors that seem to be particularly relevant in the case of risky driving are parental supervision (parents who know where their adolescent children are and how and with whom they spend their free time), parental control (parents who set and enforce rules and sanctions and have expectations for adolescents' behavior), and specific restrictions about driving (parents who set limits on when, where, and with whom adolescents are allowed to drive). Less numerous are studies which take into consideration the different roles of maternal and paternal attitudes on adolescent driving style. In general, these studies have found that both fathers and mothers are very aware of the risks that adolescents may take while driving a motor vehicle. While both parents share this opinion, mothers in particular feel that parental involvement is required, for example by setting specific restrictions on when and under what circumstances adolescents are allowed to drive (Williams et al., 2006).

As previously stated, adolescent driving style is largely influenced not only by parents, but also by peers. Friends represent a relevant model of behavior that can positively or negatively influence driving practices. Adolescents, especially boys, generally drive with friends and especially in the evening or at night (Bina, Graziano & Bonino, 2006). For adolescents, driving a car can be a way of spending time with peers, sharing experiences and emotions, and demonstrating their independence from adults and parents (Arnett, 2002). Moreover, spending time with friends who drive in a risky manner could increase the probability of adopting a risky driving style (West & Hall, 1997; Hartos et al., 2000). This can be explained with reference not only to emulation and reciprocal behavioral reinforcement within the peer group, but also to the desire to show off and surpass peers (Arnett, 2002; Bonino et al., 2005; Bina, Graziano, Borca & Calandri, 2008). In peer groups both imitation and selection of friends operate as mechanisms of influence (Ennett & Bauman, 1996; Engels & ter Bogt, 2001): imitation is not a passive conformation to specific models, but rather an active process of sharing values and behaviors within the group. Adolescents generally tend to choose and to spend time with friends who share the same ideas, attitudes, and interests (Jaccard, Dodge & Blanton, 2005). In this way adolescents define their self-image and identity through identification with significant people with whom they share their feelings (Pombeni, 1997).

The influence of friends on risky driving is linked not only to their driving model, but also to their attitudes towards risky driving. Specifically, adolescents who drive in a risky manner are more likely to have friends with a similar driving style and a similar attitude towards risky driving. Thus, even in the absence of peer pressure to take risks while driving, this kind of behavior is not only accepted, but also approved and admired by friends (Nygaard, Waiters, Grube & Keefe, 2003). Nevertheless, peer behavior and attitudes are likely to be accompanied by explicit peer pressure towards a risk behavior (Graham, Marks & Hansen, 1991). In general, adolescents who have friends who are risky drivers and who are particularly susceptible to peer pressure are more likely to drive in a risky manner (Shope, Waller & Lang, 1996). Moreover, having friends who drive in a risky way and under the influence of psychoactive substances may increase the probability of an adolescent taking risks while driving when these friends are passengers in the car. (Nygaard et al., 2003; Bonino et al., 2005). Several studies have demonstrated that when adolescents drive with others on board (especially in the evening or at night) the risk of having a road accident increases (Chen, Baker, Braver & Li, 2000; Harre, 2000; Lin & Fearn, 2003). Another aspect which has received even less attention concerns the role of passengers' gender. Boys seem to drive more safely when driving with girls, while driving with other boys increases their probability of driving in a risky manner (Arnett, 2002). Despite the fact that in western societies both males and females drive, it seems that adolescents tend to associate driving with male identity (Vavrik, 1997). As mentioned previously, and particularly when looking at the Italian context, traffic offences, especially speed limits violations, are more widespread and accepted among males than females. It is also true that, in general, sanctions are rarely applied for this type of violation (Coppo, 2004). Therefore, for some adolescents, especially boys, adopting a risky driving style is a way to prove themselves and demonstrate to others that they are strong, secure, fearless, and in control (Bonino et al., 2005).

Friendship in adolescence is characterized not only by relationships with a number of friends, but also by intimate relationships. During early adolescence friendships are generally characterized by the sharing of actions, while during middle and late adolescence the adolescent's need to share ideas, opinions, values, and personal choices with his/her friends increases. For this reason, in addition to the wider peer group, the relationship with a best friend increases in importance (Fonzi & Tani, 2000; Meeus, Branje, van der Valk & de Wied, 2007). Intimate friendships play an important adaptive role in adolescent development. However, these friendships can also be a risk factor if the best friend provides a risky behavioral model. Some research investigating this aspect has stressed the influence of the best friend on an adolescent's implication in different risk behaviors (Maxwell, 2002; Jaccard et al., 2005). Looking specifically at driving style, the best friend can be a risk factor when he/she provokes a desire to imitate, surpass, and show off. On the other hand, the regular use of safety devices (crash helmet and seat belts) by the best friend constitutes a good predictor of the same behavior by the adolescent (Bonino et al., 2005). Many studies have investigated gender differences in adolescent friendships, suggesting that relationships between boys are often characterized by the sharing of interests and activities, while relationships between girls are focused more around communication, intimacy, and confiding in one another about intimate, private experiences (Fonzi & Tani, 2000; Rose, 2002). The different impact that an intimate relationship between boys or between girls can have on an adolescent's driving practices is an aspect that has yet to be investigated and that we feel is worthy of attention. We claim, in fact, that as in peer and parent relationships, gender plays a role in intimate relationships as well.

METHOD

Aims

In line with the theoretical framework outlined, the present study has the following aims:

1. to evaluate the relationships between risky driving model and risky driving disapproval separately for father, mother, friends, and best friend;

2. to evaluate the relationships between paternal and maternal risky driving model, paternal and maternal risky driving disapproval, friends' and best friend's risky driving model, friends' and best friend's risky driving disapproval;
3. to evaluate the relationships between parents' risky driving model and risky driving disapproval (distinguishing between mother and father), friends' risky driving model and risky driving disapproval (distinguishing between friends and best friend) and adolescents' involvement in risky driving;
4. to evaluate the joint influence of all these predictors (paternal and maternal risky driving model, paternal and maternal risky driving disapproval, friends' and best friend's risky driving model, friends' and best friend's risky driving disapproval) on the adolescent's involvement in risky driving.

Gender differences were also investigated. Based on the literature illustrated previously, we expected that greater involvement of adolescents in risky driving was related to parents' and friends' risky driving models, while lower involvement was related to parents' and friends' disapproval of risky driving.

Participants

Our research involved 552 adolescents, boys and girls, ages 14 to 19, attending different high schools (lyceums, technical, and professional schools)[1] in Piedmont (northwestern Italy). Based on the aims of this particular study, a subgroup of participants was selected: adolescents who had driven at least one motor vehicle (car, motorbike, moped)[2] over the last two months met the criteria for selection. This subgroup of participants consisted of 341 adolescents drivers (50% boys), ages 14 to 19 (10% 14-15 years old, 37% 16-17 years old, 53% 18-19 years old) attending different high schools (43% lyceums, 37% technical schools, 20% professional schools). The majority of the participants had driven a car (60%)[3] or a moped (58%), while 14% had driven a motorcycle[4]. In terms of mileage, without considering the type of vehicle driven, 29% of the participants drove 1 to 10 km per week, 22% 11 to 30 km, 19% 31 to 50 km, 18% from 51 to 100 km, and finally 12% more than 100 km.

Instrument

An anonymous questionnaire, designed specifically for the aims of this study, was used to collect data. The questionnaire was administered by trained researchers in the schools to

---

[1] In Italy there are three main paths for secondary school with various specializations: lyceums (classical, scientific, linguistic, psycho-pedagogic, artistic schools), technical schools (e.g. for bookkeepers, industrial technicians, surveyors), and professional schools (e. g. for artisans, tourism operators, social service operators) (Bonino et al., 2006).

[2] By Italian law, mopeds (with piston displacements lower than 125 cc) can be driven from the age of 14 with a specific driving license, motorcycles (with piston displacements higher than 125 cc) can be driven from the age of 16 with a type A driving license, while cars can be driven from the age of 18 with a type B driving license.

[3] 43% of adolescents had driven a car without having a type B driving license. This result is in line with the findings of previous studies (Bonino et al., 2005).

[4] The total exceeds 100 because many participants had driven more than one type of motor vehicle during the last two months.

different classes during classroom time, without teachers present. Administration of the questionnaire was preceded by a presentation of the study and data were collected in accordance with Italian law and the ethical code of the Professional Psychologists Association. Completed questionnaires were turned in immediately. Statistical properties of measures used for the present study are presented.

Risky Driving

Risky driving was evaluated by considering the frequency with which adolescents' reported committing fourteen different traffic offences (e.g. exceeding the speed limit, running a red light, driving too closely to the car in front of you, driving under the effect of alcohol or marijuana) and risky driving behaviors (e.g. taking risks while driving because it makes driving more fun) over the last two months (Bonino et al., 2005). For each of the 14 items adolescents responded on a 5-point Likert scale, ranging from 0=*never* to 4= *6 or more times* (1= *once*; 2= *2-3 times*; 3= *4-5 times*) (range 0-56; N= 267; mean= 12.3; sd= 11.6; Cronbach alpha= .89).

Paternal and Maternal Risky Driving Models

The scales evaluated the adolescents' perceptions of paternal and maternal involvement in risky driving. Each scale is composed of 4 items (exceeding speed limits, committing traffic offences other than speed limit violations, taking chances while driving, driving after drinking) and adolescents responded on a 4-point Likert scale, ranging from 0= *never* to 3= *always* (1= *sometimes*; 2= *often*) (paternal risky driving model: range 0-12; N= 314; mean= 4.0; sd= 2.4; Cronbach alpha= .69; maternal risky driving model: range 0-12; N= 297; mean= 2.0; sd= 1.9; Cronbach alpha= .62).

Friends' Risky Driving Model

The scale evaluated the adolescents' perception of the number of their friends who drive in a risky manner. It is composed of 6 items (exceeding speed limits, committing traffic offences other than speed limit violations, taking chances for the fun of it when driving, driving after light drinking, driving after heavy drinking, driving after having used marijuana) and adolescents responded on a 4-point Likert scale, ranging from 0= *no friends* to 3= *all friends* (1= *some friends*; 2= *the majority of friends*) (range 0-18; N= 322; mean= 6.7; sd= 3.5; Cronbach alpha= .87).

Best Friend's Risky Driving Model

The scale evaluated the adolescents' perception of their best friend's involvement in risky driving. It is composed of 6 items (exceeding speed limits, committing traffic offences other than speed limit violations, taking chances for the fun of it when driving, driving after light drinking, driving after heavy drinking, driving after having used marijuana) and adolescents responded on a 4-point Likert scale, ranging from 0= *never* to 3= *always* (1= *sometimes*; 2= *often*) (range 0-18; N= 272; mean= 5.7; sd= 4.2; Cronbach alpha= .83).

Paternal, Maternal, Friends', and Best Friend's Disapproval of Risky Driving

The scales evaluated the adolescents' perception of paternal, maternal, friends', and best friend's disapproval of risky driving among adolescents. Each scale is composed of 4 items (exceeding speed limits, taking chances when driving, driving after light drinking, driving after heavy drinking) and adolescents responded on a 4-point Likert scale, ranging from 0= *completely approve* to 3= *completely disapprove* (1= *partly approve*; 2= *partly disapprove*) (paternal disapproval: range 0-12; N= 326; mean= 10.2; sd= 1.8; Cronbach alpha= .60; maternal disapproval: range 0-12; N= 330; mean= 10.7; sd= 1.6; Cronbach alpha= .63; friends' disapproval: range 0-12; N= 330; mean= 6.7; sd= 2.8; Cronbach alpha= .77; best friend's disapproval: range 0-12; N= 325; mean= 7.4; sd= 3.0; Cronbach alpha= .78).

RESULTS

Correlation analyses were carried out to examine the first three aims of the present study. With regard to our first aim, weak negative correlations emerged between paternal risky driving model and paternal risky driving disapproval (r= -.30 for girls and r= -.22 for boys) and between maternal risky driving model and maternal risky driving disapproval (r= -.26 for girls and r= -. 19 for boys). On the contrary, strong negative correlations emerged between friends' risky driving model and friends' risky driving disapproval (r= -.43 for girls and r= -.55 for boys) and between best friend's risky driving model and best friend's risky driving disapproval (r= -.56 for girls and r= -.55 for boys) (see Table 1). This result may suggest less coherence between parents' attitudes and behavior in terms of risky driving.

Concerning the second aim of the study, correlation coefficients showed a strong relationship between paternal and maternal disapproval of risky driving (r= .80 for girls, r= .75 for boys) and between friends' and best friend's disapproval of risky driving (r= .79 for both boys and girls). Correlations between risky driving models were also statistically significant but weaker (correlations between paternal and maternal risky driving model r= .42 for girls and r= .28 for boys; correlations between friends' and best friend's risky driving model r= .55 for girls and r= .65 for boys) (see Table 1). This result might suggest greater conformity between paternal and maternal disapproval on one hand and the wider peer group and best friend's disapproval on the other. Behavioral models, on the other hand, seem to be more differentiated.

Table 1. Correlation matrix of adolescents' risky driving, paternal, maternal, friends' and best friend's risky driving model and risky driving disapproval (girls above the diagonal, boys under the diagonal)

	Risky driving	Paternal model	Maternal model	Friends' model	Best friend's model	Paternal disapproval	Maternal disapproval	Friends' disapproval	Best friend's disapproval
Risky driving	-	.23*	.14	.27**	.36***	.02	.01	-.11	-.16
Paternal model	.38***	-	.42***	.30***	.23**	-.30***	-.12*	-.19**	-.21***
Maternal model	.22*	.28***	-	.34***	.30***	-.26***	-.26***	-.18**	-.20**
Friends' model	.46***	.22**	.25**	-	.55***	-.17*	-.15**	-.43***	-.43***
Best friend's model	.49***	.35***	.32***	.65***	-	-.15*	-.15*	-.40***	56***
Paternal disapproval	-.17*	-.22***	-.09	-.08	-.05	-	.80***	.37***	.36***
Maternal disapproval	-.01	-.01	-.19*	-.11	-.03	.75***	-	.34***	.35***
Friends' disapproval	-.38***	-.19**	-.18*	-.55***	-.46***	.27***	.17*	-	.79***
Best friend's disapproval	-.48***	-.24**	-.25**	-.50***	-.55***	.31***	.24**	.79**	-

*p<.05 ** p<.01 *** p<.001

With regard to our third aim, as expected, statistically significant positive correlations between parental and friends' risky driving models and adolescents' involvement in risky driving emerged. Statistically significant negative correlations between parental and friends' risky driving disapproval and adolescents' involvement in risky driving were also observed (see Table 1). In particular, correlations between parental and friends' risky driving models, parental and friends' risky driving disapproval, and involvement in risky driving were higher for boys. In fact, only the correlation between maternal disapproval of risky driving and involvement in risky driving was not statistically significant. For girls, only correlations between paternal (not maternal), friends', and best friend's risky driving models and involvement in risky driving were statistically significant, while correlations between paternal, maternal, friends', and best friend's risky driving disapproval and involvement in risky driving were not statistically significant. For both boys and girls the strongest correlations were observed between friends' and best friend's risky driving models and involvement in risky driving. As for parental models, only the paternal model has a statistically significant correlation with both boys' and girls' involvement in risky driving.

To evaluate the joint effects of parental and friends' risky driving models and risky driving disapproval on adolescents' involvement in risky driving (fourth aim) hierarchical multiple regression analyses were carried out. In the first model the impact of these variables on risky driving was evaluated after partialling out the effect of weekly mileage (first step) and the effects of gender and age (second step). In the last step the variables that preliminary analysis revealed to be significantly correlated with risky driving (paternal, maternal, friends', and best friend's risky driving model; paternal, friends', and best friend's disapproval) were entered as predictors variables (see Table 2).

The results showed that all variables entered in the model accounted for 43% of the variance in risky driving involvement. The variables that independently and significantly predicted risky driving were weekly mileage, gender, paternal, friends', and best friend's risky driving models. The effect of mileage accounted for 7% of the variance and gender contributed another 9%, while age was not statistically significant. Positive beta values (respectively β= .14 for mileage and β = .26 for gender) indicate a greater involvement in risky driving for people who drive more and for boys. The addition of predictors related to models and disapproval resulted in a significant improvement in explained variance (27% increase; $p<.001$). Positive beta values for paternal model (β= .22), friends' model (β= .17), and best friend's model (β= .19) indicate greater involvement in risky driving for adolescents who perceive their fathers and friends as risky drivers.

In order to investigate gender differences, separate regression analyses were carried out for boys and girls. In the first step, weekly mileage was entered, in the second age, and in the third variables that preliminary analysis showed as significantly correlated with risky driving for boys and girls respectively.

For boys, the variables entered in the model accounted for 41% of the variance in risky driving involvement (see Table 3).

Weekly mileage and age did not have a statistically significant impact on risky driving involvement. The best friend's disapproval, however, had a strong negative impact on risky driving (β= -.38), while friends' risky driving model and paternal risky driving model had a significant positive impact (β= .31 and β= .18 respectively). None of the other predictors had a statistically significant impact. This result indicates that the adolescents who are most involved in risky driving perceive less disapproval of this behavior by their best friend and perceive their friends and fathers as more involved in risky driving.

Table 2. Predictors of risky driving (hierarchical multiple regression analysis, total sample)

Step	Predictors	Final β	R^2	ΔR^2	F	p
1	Mileage	.14*	.07	.07***	14.36	.000
2	Gender Age	.26*** .06	.16	.09***	12.72	.000
3	Paternal model Maternal model Friends' model Best friend's model Paternal disapproval Friends' disapproval Best friend's disapproval	.22** -.04 .17* .19* .10 -.04 -.15	.43	.27***	14.62	.000

* p <.05 ** p < .01 ***p<.001

Table 3. Predictors of risky driving (hierarchical multiple regression analysis, boys)

Step	Predictors	*Final β*	R^2	Δ R^2	F	p
1	Mileage	.17	.02	.02	2.96	.089
2	Age	-.07	.02	.00	1.58	.211
3	Paternal model	.18*				
	Maternal model	-.11				
	Friends' model	.31**				
	Best friend's model	.10	.42	.40***	8.60	.000
	Paternal disapproval	.13				
	Friends' disapproval	.05				
	Best friend's disapproval	-.38*				

* p <.05 ** p < .01 ***p<.001

For girls, the variables entered in the model accounted for 27% of the variance in risky driving involvement (see Table 4).

Table 4. Predictors of risky driving (hierarchical multiple regression analysis, girls)

Step	Predictors	*Final β*	R^2	Δ R^2	F	p
1	Mileage	.19	.07	.07**	7.75	.007
2	Age	.24*	.09	.02	5.69	.005
3	Paternal model	.16				
	Friends' model	.11	.27	.18***	7.64	.000
	Best friend's model	.30**				

* p <.05 ** p < .01 ***p<.001

Weekly mileage did not have a statistically significant impact on risky driving involvement. The variables that independently and significantly predicted risky driving were age and best friend's risky driving model. The effect of age accounted for 9% of the variance. A positive beta value (β= .24) indicates greater involvement in risky driving for older girls. The addition of predictors related to models resulted in a significant improvement in explained variance (increment 18%; p<.001). Only the best friend's risky driving model had a statistically significant impact on risky driving (β= .30), while paternal and friends' models were not statistically significant. This result indicates a greater involvement in risky driving for girls who perceive their best friend as a risky driver, regardless of the perceived driving model of their father and their broader group of friends.

CONCLUSION

The present study stressed the important influence of the driving model offered by parents and friends as well as their disapproval of risky driving on risky driving in adolescence. As we expected, greater involvement of adolescents in risky driving was related to parents' and friends' risky driving model, while lower involvement was related to parents' and friends' disapproval of risky driving. These results are consistent with the findings of previous studies which indicate that both parents' and friends' behavior (Bianchi & Summala, 2004; Bonino et al., 2005; Wilson et al., 2006) and attitudes (Taubman-Ben-Ari et al., 2005; Williams et al., 2006) towards risk behavior can influence adolescents' implication in the same behavior. Moreover, this study, considering parents' and friends' risky driving model and disapproval simultaneously, showed that adolescents' involvement in risky driving was more closely related to parents' and friends' risky driving behavior than to their disapproval of risky driving.

With regard to the relationship between driving model and disapproval of risky driving in parents and friends, our results stressed a general coherence between risky driving behavior and attitude. In other words, in the adolescents' perception, a parent or a friend who drives in a risky way demonstrates less disapproval of risky driving. In fact, correlations between maternal, paternal, friends', and best friend's models and disapproval of risky driving were negative and statistically significant. Nevertheless, correlations between behavior (risky driving model) and attitude (risky driving disapproval) were lower for parents than for friends. This result reveals that adolescents may not perceive the same strong coherence between how their parents drive and what they think and say about risky driving that they perceive in their friends. Previous research has found that parental risky driving disapproval plays a protective role especially when it is coherent with parental driving behavior; in other words, when parents are safe drivers (Bonino et al., 2005). It appears that parental disapproval of risky driving alone is not enough to prevent adolescent involvement in risky driving; it is important for parents to provide a positive driving model as well. Our results also showed that parental disapproval was less related to risky driving than to parental risky driving behavior (only paternal disapproval was correlated to boys' involvement in risky driving).

Comparing behavioral models and attitudes of mothers and fathers, and of friends and best friends, correlations showed a strong relationship between paternal and maternal disapproval of risky driving and between friends' and best friend's disapproval. On the contrary correlations between risky driving models were statistically significant but weaker. This result might suggest greater conformity in the attitudes of fathers and mothers, as well as in the attitudes of the wider peer group and the best friend. Behavioral models, however, seem to be more differentiated.

Some interesting results derived from analyzing separately the relationships between risky driving and each of the predictors. In terms of parental risky driving models, comparing the role of fathers and mothers, we found that for both boys and girls, the paternal risky driving model was more related to adolescents' involvement in risky driving than the maternal model, while maternal risky driving behavior was related to greater involvement in risky driving for boys only. This result reflects the findings of other recent studies suggesting the significant influence of the paternal model on sons' and daughters' driving behaviors (Taubman-Ben-Ari et al., 2005; Bonino et al., 2005). The great importance of perceived paternal driving behavior is probably tied to the fact that driving and risky driving (Vavrik, 1997; Bina et al., 2006; Møller, 2004), particularly in Italy, are considered typically masculine behaviors. In light of this, boys in particular may be more sensitive to the paternal driving model than to the maternal model. As mentioned earlier, some Italian research has suggested that paternal driving behavior can also play an important protective role: for example, when fathers habitually use safety devices, it is more likely that their sons and daughters will do so as well (Bonino et al., 2005).

Considering friends' models and disapproval, our results showed relevant gender differences. First of all, risky driving disapproval by both friends and best friend were not significantly correlated to involvement in risky driving for girls; only friends' and best friend's behavior played a significant role. On the contrary, both the risky driving model and the disapproval of friends and the best friend were strongly related to risky driving for boys. These results suggest a stronger tendency by boys to comply with their friends, in terms of both attitudes and behaviors.

Further evidence emerged investigating the combined action of both parents' and friends' risky driving model and disapproval. Multiple regression analysis indicated that, in general, friends and best friends play a more relevant role than parents in adolescents' risky driving. In this regard, it is widely known that, while family relationships remain significant, peers gain more and more importance during adolescence. Parents and friends serve as important points of reference in different areas; while adolescents generally consult their parents on important matters, such as decisions regarding their scholastic or professional careers, they usually refer to their friends when it comes to their looks, leisure activities, and other daily interests (Coleman & Hendry, 1990; Zani, 1997). Driving behavior also seems to be strongly affected by the influence of friends. Moreover, a group of friends may represent a context where different behaviors, sometimes in opposition to parents' models, are acted out; thus risky driving may be a form of transgression against social and family rules (Bonino et al. 2005; Bina et al. 2006; Williams, 1998). In this case, even if parents aren't risky drivers and disapprove of this behavior, adolescents might use risky driving to gain the approval of their friends and to follow their friends' behavior models.

Relevant gender differences emerged in this area as well. For boys, both friends (especially friends' model) and best friend (especially best friend's disapproval) played a role in risky driving. For girls, the best friend's model was the only significant predictor of involvement in risky driving. Thus, girls' risky driving is unrelated to friends' model and disapproval of risky driving. Researchers have stressed that adolescents and their friends share similar ideas and behaviors and that this similarity is the result of two different processes: friends' influence on one hand and the selection of friendships on the other (Ennett & Bauman, 1996; Rutter, Giller &Hagell 1998; Engels & ter Bogt, 2001). The fact that girls' driving behavior is similar to the best friend's model, but not to the model of the larger group of friends, could be related to the different manner in which boys and girls form intimate friendships. Research has shown that girls usually have fewer friends than boys and that girls' relationships tend to be characterized more by trust, intimacy, and emotional closeness (Collins & Repinski, 1994). On the contrary, boys form friendships based on shared activities with a larger number of friends (Caprara & Fonzi, 2000; Rose, 2002). Some scholars have claimed that differences in boys' and girls' friendships reflect different ways to validate their self worth (Buhrmester & Furman, 1993; Furman, 1996). In their group of friends, boys find support in the process of constructing their identity by conforming to the attitudes and behaviors of the group, especially for minors. In keeping with this previous finding, our results pointed to conformity between adolescents' risky driving behavior and their friends' behavior for boys only. On the contrary, it seems that for girls, the behavior of their best friend and their larger group of friends does not play a relevant role in risky driving.

While the role of friends seems to be more important, our results showed that paternal driving must also be taken into account, particularly in the case of boys. Considering gender differences, only boys who perceived their fathers as risky drivers were more involved in risky driving. As previously noted, our results suggest that paternal driving behavior, viewed as a typically masculine behavior, can be especially influential for their sons' driving. Young male drivers are involved in the process of constructing their adult and masculine identity; driving and risky driving are functional to this process and their fathers' model plays a relevant role for boys. Our study indicates that mothers' driving behavior seems unrelated to adolescents' risky driving, even for girls. This is likely due to the fact that society (especially Italian society) does not consider driving a typically feminine behavior.

Our study has some limitations. First of all, the measures were self-reported. Nonetheless, research on the reliability of self-reported driving behavior has shown that it is generally reliable (Hatakka et al., 1997; Lajunen & Summala, 2003). With regard to parents' and friends' risky driving model and disapproval of risky driving, while we recognize the value of investigating these aspects by asking parents and friends directly about their behavior and attitudes, from a constructivistic point of view, it has been repeatedly demonstrated that the individual's perception of reality is an important source of information in understanding individual action (Magnusson & Stattin, 1997; Juang & Silbereisen, 2001). A further limitation is linked to the fact that the cross-sectional design of our research does not allow a causal interpretation of the data. We recognize the need to develop a longitudinal research design in order to better investigate the potential risk or protective role of parental and friends' behavior and attitudes with respect to adolescent involvement in risky driving. Finally, caution should be used when generalizing these findings to adolescents with different demographic characteristics or from different countries.

Despite these limitations, results from our study provided some important suggestions for risky driving prevention programs targeting adolescent drivers. First of all, effective interventions should take into account the relevant role of friends with respect to adolescent involvement in risky driving. Moreover, the role of parents, in particular fathers, should not be neglected. Interventions involving parents should stress that a parental model of safe driving is necessary to protect adolescents from serious involvement in risky driving and that disapproval of risky driving behavior does not suffice if parental driving behavior is not coherent with this attitude. Finally, in light of the gender differences revealed by our research, it may be useful to design different interventions for boys and girls.

ACKNOWLEDGMENTS

The authors are grateful to prof. Silvia Bonino for her precious supervision of the entire research project, to Valentina Aprile for her contribution to data collecting, data entry and bibliographic research and to Lindsay McDonald for her English revision of the manuscript.

REFERENCES

Arnett, J. J. (1996). Sensation seeking, aggressiveness, and adolescent reckless behavior. *Personality and Individual Differences, 20*, 693-702.

Arnett, J. J. (2002). Developmental sources of crash risk in young drivers. *Injury Prevention, 8*, 17-23.

Arnett, J. J., Offer, D. & Fine, M. A. (1997). Reckless driving in adolescence: "state" and "trait" factors. *Accident Analysis and Prevention, 29*, 57-63.

Asbridge, M., Poulin, C. & Donato A. (2005). Motor vehicle collision risk and driving under the influence of cannabis: evidence from adolescents in Atlantic Canada. *Accident Analysis and Prevention, 37*, 1024- 1034.

Beirness, D. J. (1993). Do we really drive as we live? The role of personality factors in road crashes. *Alcohol Drugs Driving, 9*, 129-143.

Bianchi, A. & Summala, H. (2004). The "genetics" of driving behavior: parents' driving style predicts their children's driving style. *Accident Analysis and Prevention, 36*, 655-659.

Bina, M., Graziano, F. & Bonino, S. (2006). Risky driving and lifestyle in adolescence. *Accident analysis and Prevention, 38*, 472-481.

Bina, M., Graziano, F., Borca, G. & Calandri, E. (2008). The psychological functions of risky driving in adolescence: the adolescents' point of view. In A. De Smet (Ed.), *Transportation Accident Analysis and Prevention* (pp. 1-24). New York: Nova Science Publishers Inc.

Bonino, S., Cattelino, E. & Ciairano, S. (2005). *Adolescents and risk. Behaviors, functions and protective factors*. New York: Springer Verlag.

Bonino, S., Cattelino, E. & Ciairano, S. (2006). Italy. In J.J. Arnett (Ed.) *Routledge International Encyclopedia of Adolescence* (pp.510-523). New York: Routledge.

Buhrmester, D. and Furman, W. (1993), Age and sex differences in perceptions of networks of personal relationships. In S.S. Feldman and G.R. Elliott, (Eds) *At the Threshold: The Developing Adolescent*. Massachusetts: Harvard University Press.

Caprara, G. V. & Fonzi, A. (Eds.) (2000). *L'età sospesa* [The uncertain age]. Firenze: Giunti.

Chen, L. H., Baker, S. P., Braver, E. R. & Li, G. (2000). Carrying passengers as a risk factor for crashes fatal to 16- and 17- year-old drivers. *Journal of the American Medical Association, 283*, 1578-1582.

Coleman, J. & Hendry L. (1990). *The nature of adolescence.* London: Methuen.

Collins, W. A. & Repinski, D. J. (1994). Relationships during adolescence: Continuity and change in interpersonal perspective. In R. Montemayor, G. R. Adams, & T.P . Gullotta (Eds.), *Personal relationships during adolescence* (pp. 7-36). Thousand Oaks: Sage.

Coppo, M. (2004). *Rapporto della Consulta Nazionale per la Sicurezza Stradale*, Ministero degli Interni, Italia [*Report of the Road Safety National Council*, Ministry of Internal Affairs, Italy].

Deffenbacher, J. L., Lynch, R. S., Oetting, E. R. & Yingling, D.A. (2001). Driving anger: correlates and test of state-trait theory. *Personality and Individual Differences, 31*, 1321-1331.

Engels, R. & ter Bogt, T. (2001). Influences of risk behaviors on the quality of peer relations in adolescence. *Journal of Youth and Adolescence, 30*, 675-695.

Ennett, S. T. & Bauman, K. E. (1996). Adolescent social networks: School, demographic and longitudinal considerations. *Journal of Adolescent Research, 11*(2), 194–215.

ERSO (European Road Safety Observatory) (2008). *Annual Statistical Report.* http://www.erso.eu

Ferguson, S. A., Williams, A. F., Chapline, J. F., Reinfurt, D. W. & De Leonardis, D.M. (2001). Relationship of parent driving records of their children. *Accident Analysis and Prevention, 33*, 229-234.

Finken, L. L., Jacobs, J. E. & Laguna, K. D. (1998). Risky driving and driving/riding decisions: the role of previous experience. *Journal of Youth and adolescence,27* (4), 493-511.

Fonzi, A. & Tani, F. (2000). Amici per la pelle. Le caratteristiche dei legami amicali nell'adolescenza [Bosom friends. Characteristics of friendship in adolescence]. In G.V. Caprara & A. Fonzi (Ed.), *L'età sospesa [The uncertain age](*pp. 90-120). Firenze: Giunti.

Furman, W. (1996) The mesurement of friendships perception: conceptual and methodological issues. In W.M. Bukowski, A.F. Newcomb, W.W. Hartup (Eds.) *The Company They Keep: Friendship in Childhood and Adolescence,*Cambridge University Press, New York.

Graham, J. W., Marks, G. & Hansen, W.B. (1991). Social influence processes affecting adolescent substance use. *Journal of Applied Psychology, 79*, 291-298.

Harre, N. (2000). Risk evaluation, driving, and adolescents: a typology. *Developmental Review, 20*, 206-226.

Harris, K. M. & Morgan, S. P. (1991). Fathers, sons, and daughter: Differential paternal involvement in parenting. *Journal of Marriage and the Family, 53*, 531-544.

Hartos, J., Eitel, P., Haynie, D. L. & Simons-Morton, B. (2000). Can I take the car? Relations among parenting practices and adolescent problem-driving practices. *Journal of Adolescent Research, 15* (3), 352-367.

Hartos, J., Eitel, P. & Simons-Morton, B. (2002). Parenting practice and adolescent risky driving: a three-month prospective study. *Health Education and Behavior, 29*, 194-206.

Hatakka, M., Keskinen, E., Gregersen, N. P., Glad, A. & Hernetkoski, K. (2002). From control of the vehicle to personal self-control; broadening the perspectives to driver education. *Transportation Research* Part F, 5, 201–15.

Hatakka, M., Keskinen, E., Katila, A. & Laapotti, S. (1997). Self-reported driving habits are valid predictors of violations and accidents. In T. Rothengatter & E. Carbonell Vaya (Eds.), *Traffic and transport psychology. Theory and application* (pp. 295-303). Pergamon.

ISTAT (National Institute of Statistics)/ACI (Automobile Club Italy) (2007). *Incidenti stradali. Anno 2005 [Traffic accidents statistics, 2005].* Rome: ISTAT.

Jaccard, J., Dodge, T. & Blanton, H. (2005). Peer influence on risk behavior: an analysis of the effects of a close friend. *Developmental Psychology, 41* (1), 135-147.

Jessor, R. (1987). Risky driving and adolescent problem behavior: an extension of problem-behavior theory. *Alcohol, Drugs and Driving, 3*, 1–11.

Jessor, R. (Ed.) (1998). *New Perspectives on Adolescent Risk Behaviour.* Cambridge: Cambridge University Press.

Jessor, R., Donovan, J. E. & Costa, F. M. (1991). *Beyond Adolescence. Problem Behavior and Young Adult Development.* New York: Cambridge University Press.

Jessor, R., Turbin, M. S. & Costa, F. M. (1997). Predicting developmental change in risky driving: the transition to young adulthood. *Applied Developmental Science, 1* (1), 4-16.

Jonah, B. A. (1986). Accident risk and risk taking behavior among young drivers. *Accident Analysis and Prevention, 18*, 255-271.

Jonah, B. A. (1990). Age differences in risky driving. *Health Education Research, 5*, 139-149.

Juang, L. P. & Silbereisen, R. K. (2001). Family transitions for young adult women in the context of a changed Germany: timing, sequence and duration. *American Behavioral Scientist, 44*, 1899-1917.

Kandel, D. B. & Andrews, K. (1987). Process of adolescent socialization by parents and peers. *International Journal of the Addictions, 22*, 319-342.

Koopmans, J. R., Boomsma, D. I., Heath, A. C. & Vandoornen, L.J.P. (1995). A multivariate genetic-analysis of sensation seeking. *Behavior Genetics, 25* (4), 349-356.

Laapotti, S., Keskinen, E., Hatakka, M. & Katila, A. (2001). Novice drivers' accidents and violations – a failure on higher or lower hierarchical levels of driving behaviour. *Accident Analysis and Prevention, 33*, 759-769.

Lajunen, T. & Summala, H. (2003). Can we trust self-reports of driving? Effect of impression management on driving behaviour questionnaire responses. *Transportation Research*, F-6, 97-101.

Lin, M. L. & Fearn, K. T. (2003). The provisional license: nightime and passenger restrictions – a literature review. *Journal of Safety Research, 34*, 51-61.

Magnusson, D. & Stattin H. (1997). Person-context interaction theories. In R. Lerner & W. Damon (Eds.), *Theoretical models of human development, vol. 1, Handbook of child psychology.* New York: Wiley.

Maxwell, K. A. (2002). Friends: the role of peer influence across adolescent risk behaviors. *Journal of Youth and Adolescence, 31* (4), 267-277.

Meeus, W. H. J., Branje, S. J. T., van der Valk, I. & de Wied, M. (2007). Relationships with intimate partner, best friend and parents in adolescence and early adulthood: A study of the saliency of the intimate partnership. *International Journal of Behavioral Development, 31* (6), 569-580.

Møller, M. (2004). An explorative study of the relationship between lifestyle and driving behaviour among young drivers. *Accident Analysis and Prevention, 36*, 1081-1088.

Møller, M. & Gregersen, N. P. (2008). Psychosocial function of driving as predictor of risk-raking behaviour. *Accident Analysis and Prevention, 40*, 209-215.

Nygaard, P., Waiters, E. D., Grube, J. W. & Keefe, D. (2003). Why do they do it? A qualitative study of adolescent drinking and driving. *Substance Use and Misuse, 38* (7), 835-863.

Nygaard, P. & Grube, J.W., (2005). Mixed messages: Contributions to adolescent drinking and driving. *Addiction Research and Theory, 13* (5), 411-426.

Pombeni, M. L. (1997). L'adolescente e i gruppi di coetanei [Adolescents and peer groups]. In A. Palmonari (Ed.), *Psicologia dell'adolescenza [Psychology of adolescence]*(pp. 251-270). Bologna: Il Mulino.

Rose, A. J. (2002). Co-rumination in the friendships of girls and boys. *Child Development, 73*, 1830-1843.

Rutter, M., Giller, H. & Hagell, A. (1998). *Antisocial behavior by young people.* New York: Cambridge University Press.

Shope, J. T., Waller, P. F. & Lang, S. W. (1996). Alcohol-related predictors of adolescent driving: gender differences in crashes and offences. *Accident Analysis and Prevention, 28 (6),* 755-764.

Shope, J. T., Raghunathan, T. E. & Patil, S. M. (2003). Examining trajectories of adolescent risk factors as predictors of subsequent high-risk driving behaviour. *Journal of Adolescent Health, 32,* 214-224.

Taubman-Ben-Ari, O., Mikulincer, M. & Gillath, O. (2005). From parents to children – similarity in parents and offspring driving styles. *Transportation Research Part F, 8,* 19-29.

Vavrik, J. (1997). Personality and risk-taking: A brief report on adolescent male drivers. *Journal of Adolescence, 20,* 461-465.

West, R. & Hall, J. (1997). The role of personality and attitudes in traffic accident risk. *Applied Psychology: An International Review, 46,* 253-264.

Williams, A.F. (1998). Risky driving behaviour among adolescents. In R. Jessor (Ed.), *New perspectives on adolescent risk behaviour* (pp. 221-237). Cambridge: Cambridge University Press.

Williams, A. F. (2003). Teenage drivers: patterns of risk. *Journal of Safety Research, 34,* 5-15.

Yu, J. & Williford, W. R. (1993). Alcohol and risk sensation seeking. Specifying a causal model on high-risk driving. *Journal of Addictive Diseases, 12,* 79-86.

Williams, A. F., Leaf, W. A., Simons-Morton, B. G. & Hartos, J. L. (2006). Parents' views of teen driving risks, the role of parents, and how they plan to manage the risks. *Journal of Safety Research, 37,* 221-226.

Wilson, R. J., Meckle, W., Wiggins, S. & Cooper, P.J. (2006). Young driver risk in relation to parents' retrospective driving record. *Journal of Safety Research, 37,* 325-332.

Zani B. (1997). L'adolescente e la sessualità [Adolescents and sexuality]. In A. Palmonari (Ed.), *Psicologia dell'adolescenza [Psychology of adolescence]*(pp. 177-202). Bologna (Italy): Il Mulino.

Zuckerman, M. (1993). Impulsive sensation seeking and its behavioral psychophysiological and biochemical correlates. *Neuropsychobiology, 28* (1-2), 30-36.

In: Accidents: Causes, Analysis and Prevention
Editors: Henri Bédard and Geraud Delashmit
ISBN: 978-1-60741-712-5

Chapter 4

CHARACTERISTICS AND CRASH RELATED BEHAVIOUR OF VULNERABLE ROAD USERS PRESENTED TO EMERGENCY DEPARTMENTS

Lynn B. Meuleners, Andy H. Lee* and Michelle L. Fraser
Centre for Population Health Research, School of Public Health,
Curtin University of Technology, Perth, Australia

ABSTRACT

Vulnerable road users, namely, motorcyclists, bicyclists and pedestrians, contribute significantly to morbidity and mortality rates for road casualties. Many of these injuries occur off-road without police attendance or do not involve a collision with a motor vehicle. As a result, the magnitude and scope of their injuries are often underestimated. Behavioural factors also play a pivotal role, but information on the prevalence of behaviours related to an increased risk of involvement in a crash is also rather limited.

A six-month prospective study of vulnerable road users presenting to emergency departments due to an injury from a crash or non-motor vehicle collision was conducted in Perth, Western Australia. Through a structured questionnaire, the characteristics of the road user, the injury sustained and the crash were described and the crash-related behaviour investigated.

Of the total 392 vulnerable road users, 282 (72%) participated in the interview with the majority being bicyclists (53%) followed by motorcyclists (37%). Males represented 80% of the presentations and dominated each road user type except pedestrians. Many bicyclists and motorcyclists were injured by a collision that did not involve a motor vehicle and/or occurred in off-road situations and was not attended by police. In terms of behavioural factors, 30% of pedestrians reported the motor vehicle involved was moving faster than initially thought, particularly among older adults aged 55 and above. The study found that 7% of motorcyclists and 23% of bicyclists were not wearing their helmet at all times, while one third of the adolescent bicyclists did not wear a helmet at the time of injury. Only 20% of bicyclists and 67% of motorcyclists injured during the evening

* Correspondence author: Professor Andy H. Lee, School of Public Health, Curtin University of Technology,GPO Box U 1987, Perth, WA, 6845, Australia, Phone: +61-8-92664180, Fax: +61-8-92662958, Email: Andy.Lee@curtin.edu.au

were wearing visible clothing. Overall, 14% of vulnerable road users admitted being distracted and 13% of them were hurrying before the injury event. Excessive speed was a contributing factor in crashes sustained by motorcyclists.

In conclusion, data linkage between emergency presentations, hospital admissions and police records is recommended to improve the accuracy and representativeness of injury data. Behavioural factors play a pivotal role in the protection of vulnerable road users. Their safety, to a large extent, is determined by their risk taking behaviour. Safety promotion programs targeting the wearing of helmet and visible clothing should be aimed at both motorcyclists and bicyclists, especially among adolescent bicyclists.

1. Introduction

Vulnerable road users, defined as motorized two wheeler riders (motorcyclists), bicyclists and pedestrians, contribute significantly to morbidity and mortality rates due to road crashes [1,2]. During 2006 in the state of Western Australia, 27% of people killed and 23% of those hospitalised in road crashes were motorcyclists, bicyclists and pedestrians [3]. For these vulnerable road users, injury can be caused by a collision with a motor vehicle, a fall, or a collision with a fixed object, person or animal. In this chapter, the term "crash" will refer to any of these events.

To date, the majority of information on vulnerable road user crashes has been obtained from police reports. However, studies have shown that these crashes are underreported, especially when they do not involve a collision with a motor vehicle or do not occur on a public roadway [4-6]. Consequently, information obtained from police data does not accurately represent the magnitude and scope of vulnerable road user crashes.

A prospective, descriptive study conducted in Western Australia of motorcyclists, bicyclists and pedestrians who presented to hospital emergency department (ED) as the result of a crash has provided valuable information on the demographic, crash, behavioural and injury characteristics of these vulnerable road users [7]. An understanding of the range of characteristics associated with vulnerable road user crashes will provide a basis for the development of effective counter measures for injury prevention among this group.

This chapter examines the injury risks experienced by vulnerable road users and challenges in obtaining accurate information on the magnitude and circumstances of their crashes. In addition, previous findings on motorcyclist, bicyclist and pedestrian crashes are reviewed in terms of demographic and injury patterns, as well as environmental factors and behavioural characteristics. The Western Australian study of ED presentations by vulnerable road users is next described in detail. Finally, implications of the study on road safety and recommendations for future research are discussed.

2. Injury Risks for Vulnerable Road Users

Every year, approximately 1.2 million people are killed in road traffic crashes worldwide and as many as 50 million are injured or disabled [1]. These numbers are predicted to increase by 65% between the years 2000 and 2020 [1]. By 2020, it is estimated that traffic crashes will be the third leading cause of death and disability worldwide after heart disease and depression [8]. The direct global economic cost arising from road crashes is considerable and has been estimated at US$ 518 billion annually [1].

Vulnerable road users account for a significant proportion of crash-related morbidity and mortality [1,3]. In low income countries, the majority of road mortality is experienced by vulnerable road users whereas car occupants are the leading casualties in high income countries. Although there are relatively less motorcyclist, bicyclist and pedestrian casualties in high income countries, when fatality rates based on measures of exposure are compared for all road users, the risk of dying in a road crash is much higher for vulnerable road users almost everywhere in the world [1]. A recent study calculated exposure-based road injury rates in the United States and reported that when all road users were compared, fatal injury rates were highest for motorcyclists, pedestrians and bicyclists and non-fatal injury rates highest for motorcyclists and bicyclists [9]. These findings mirror those from European countries [10].

Pedestrians, bicyclists and motorcyclists experience an increased risk of crashing and an increased risk of injury in the event of a crash because of their vulnerability [10]. The increased risk of injury is due to the 'unprotected' state of vulnerable road users as opposed to 'protected' motor vehicle occupants [1,11]. When compared with motor vehicle occupants, pedestrians and bicyclists have been reported to sustain higher rates of head injuries, such as skull fractures, epidural haemorrhage, subdural haemorrhage and brain contusion, as well as injuries to the lower extremities [11].

3. Challenges Obtaining Information on Vulnerable Road User Crashes

Traditionally, police reports have provided the primary source of information on crashes causing injury to road users. Past research has demonstrated extensive under-reporting of crashes to police for vulnerable road users [4-6,12]. For example, a recent assessment of linked data in New South Wales, Australia, reported that only 69% of road traffic crashes from hospital records had matching police crash records [5]. The odds of linkage for pedestrians, motorcyclists and bicyclists were only 0.47, 0.29 and 0.14 that of motor vehicle drivers, respectively [5]. Similarly, an analysis of linked police and hospital data relating to bicycle crashes in Western Australia from 1987-2000 found that bicyclists accounted for almost 14% of road casualties admitted to hospital but only 1% of crash involvement according to police records [12].

One explanation for the extensive underestimation of crashes by vulnerable road users is that off road crashes, falls and collisions not involving a motor vehicle are rarely reported to the police [13,14]. Police records are mostly limited to motor vehicle related crashes occurring on public roads that result in either injury or property damage exceeding a certain value. They commonly exclude crashes that occur in non-roadway locations such as parking lots, driveways, footpaths and cycle paths. They may also exclude non-collision events that do not involve a motor vehicle such as falling from a bicycle. This is of concern because past research has shown that non-motor vehicle and non-roadway crashes pose significant injury risks for vulnerable road users [14].

Hospital morbidity data can provide another source of valuable information about injured vulnerable road users but also have weaknesses. These records are generally restricted to patients actually admitted to hospital and therefore only capture those severely injured. Although they provide in-depth detail on the injury, little information is available concerning the crash event. Data from hospital ED are emerging as a potentially important source of information, especially for capturing less severely injured vulnerable road users involved in falls or non-motor vehicle collisions.

4. Vulnerable Road User Crashes: Evidence to Date

Vulnerable road users experience increased risk of injury and death due to crashes. Relative to passenger vehicle occupants, motorcyclists, bicyclists and pedestrians are 58.3, 2.3 and 1.5 times respectively, more likely to be fatally injured on a given trip [9]. Characteristics associated with a crash can be separated into demographic/ injury factors, environmental and behavioural factors. A thorough understanding of the characteristics associated with each vulnerable road user group is essential for developing relevant interventions and policies to reduce crashes and injury severity. The available evidence on motorcyclist, bicyclist and pedestrian crashes will be discussed separately because the circumstances underlying their crashes are different.

4.1. Motorcyclists

After a period of sustained reduction, motorcyclist fatalities and injuries appear to be increasing in Australia. In Western Australia, motorcycle fatalities and hospitalisations were 33% and 18% higher than the respective previous five year average [3]. Accurate information on the size and nature of the issue cannot be obtained from police records only. Single vehicle motorcycle crashes as well as crashes involving unlicensed riders, unregistered motorcycles or illegal blood alcohol levels are unlikely to be reported to police unless they are very severe [15]. Consequently, the number of motorcycle crashes in which these factors are involved will be underestimated.

4.1.1. Demographic and injury patterns of motorcycle crashes

Motorcycle riding is predominantly a male sport, which may account for the over-representation of males in motorcycle crashes. Over eight times the number of male motorcyclists were killed or hospitalised in Western Australia in 2006 than females [3]. A Swedish cohort study also reported a risk of injury 2.5 times higher among motorcyclists aged 17-19 years of low socio-economic background than the highest socio-economic group [16]. Motorcyclists often sustain more severe injuries than other road users. A previous study found that 4% of all injured motorcyclists had severe head injuries, 44% had upper limb injuries and 57% had lower limb injuries [17].

4.1.2. Environmental characteristics of motorcycle crashes

A variety of environmental factors affecting motorcycle crash or injury have been identified. It is known that 80% of crashes occurred in urban areas and almost half on major arterial roads [17]. Sixty seven percent of crashes occurred when the weather was fine and over 40% took place on roads with no street lights [17]. More recently, studies have identified environmental and crash factors associated with injury severity. These include riding in a rural area, areas with a speed limit over 50 km per hour, poor visibility, right angle and head on collisions and collisions with fixed objects [18,19].

4.1.3. Behavioural characteristics of motorcycle crashes

Much of the investigation into behavioural risk factors for motorcyclists has focused on helmet use and alcohol consumption. A Cochrane review of the literature estimated that motorcycle helmets can reduce the risk of death by 42% and the risk of head injury by 69% [20]. Alcohol consumption has been associated with both greater crash risk and more severe injuries [18,19,21]. Meanwhile, speed has been identified as a key contributor to motorcycle crashes and injury severity [18,21]. Speed is especially related to fatal crashes among riders under the age of 30 years [22]. Interestingly, wearing appropriate clothing did not significantly reduce the likelihood or severity of external injuries in the event of a motorcycle crash [17].

A recent study in the United Kingdom developed a questionnaire that measures the behaviour of motorcyclists and tested which behaviours could predict crash risk [23].Traffic errors (slips, lapses and mistakes) were found to be the main predictors of crash risk, whereas control errors and speed violations were significant predictors of crashes when the respondent accepted some degree of blame [23].

4.2. Bicyclists

Bicycle riding is a popular recreational activity and mode of transportation worldwide. A large body of literature already exists on risk factors for bicyclist crashes, injuries and deaths. In Western Australia, bicyclists account for almost 14% of road users admitted to hospital due to injury [12].

4.2.1. Demographic and injury patterns of bicycle crashes

Children, adolescents and older adults over 55 years pose the greatest risk of being killed or injured due to a bicycle crash [24-26]. Amongst four Australian states, riders aged 6-19 years were involved in most crashes [26]. Interestingly, hospitalised cyclists in Western Australia were considerably younger than cyclists from police-reported data [12].

Bicycle crash, injury and fatality rates have been consistently higher for males than females [9,26,27], with male cyclists making up 80% of police reported bicycle crashes in four Australian states [26]. Moreover, males had a greater exposure rate and also exhibited more risk taking behaviours than females [28]. Injury risk to child and adolescent bicyclists was found to be greater among those from lower socio-economic groups in Sweden [16].

In Western Australia, bicyclists suffered fewer critical, severe and serious injuries (13%) than pedestrians (32%), motorcyclists (26%) and motor vehicle occupants (26%), but the highest proportion of moderate injuries (65%) [12]. For cyclists, upper extremity injuries were most frequent but head injuries accounted for the highest proportion of critical injuries [12]. Nearly 50% of children with bicycle related injuries suffered injuries to the neck and head [29]. However, a major limitation of such studies is the lack of exposure profiles among different age groups.

4.2.2. Environmental characteristics of bicycle crashes

Riding in inclement weather, in darkness with no streetlights or during morning peak hour all doubled the probability of a bicyclist experiencing a fatal injury in a bicycle-motor vehicle crash [24]. Colliding head on, with a speeding motor vehicle or with an intoxicated motor vehicle driver also more than doubled the risk of fatality [24]. An analysis of ED presentations by cyclists confirmed that night-time cycling was two to five times more dangerous than cycling in the daylight with 40% of bicyclist fatalities occurred during darkness [14]. Furthermore, injury severity of bicyclists increased after dark as well as in fog, in areas with higher speed limits and on roads with and an up/ downgrade [30].

In terms of the relative rates for falls and injuries, cycling on-road may be safer than cycling off-road on footpaths, cycle paths or tracks [31]. A study of ED presentations found that 70% of bicycle injuries did not involve a motor vehicle and 31% were injured in non-roadway locations such as parking lots, pathways or off-road trails but that these crashes were less likely to require hospitalisation [14]. Similarly, 82% of bicyclist crashes requiring hospitalisation did not involve a motor vehicle and that only 40% occurred on a public road [12]. Moreover, among children and adolescents, 47.5% of bicyclist injuries occurred at home[27].

4.2.3. Behavioural characteristics of bicycle crashes

Wearing a bicycle helmet has been shown to protect against head and brain injury [32] and to reduce the probability of fatal or severe injuries in the event of a crash [33,34]. Alcohol consumption has been associated with more than double the probability of fatal injury [24]. When compared with sober bicyclists, intoxicated bicyclists had a greater risk of injury to the head or face, of being involved in night time and single vehicle crashes and less often wore a helmet [35]. Meanwhile, bicycle safety equipment, other than brakes and reflectors, were used by less than 7% of children presenting to ED as the result of a crash [29].

4.3. Pedestrians

In Western Australia, pedestrians represented 7% of road users killed or hospitalised during 2006 [3]. Crashes involving pedestrians produce particularly more serious injuries than other road users [36].

4.3.1. Demographic and injury patterns of pedestrian crashes

Children and elderly pedestrians appear to be the age groups most commonly injured in road crashes while more male pedestrians are affected than females [36]. Pedestrians aged 60 years and older accounted for 40% of all pedestrian fatalities in 1995 even though they made up only 15% of the population [37]. In addition, once involved in a crash, injury to an older pedestrian can be more severe. Indeed, the proportion of fatal injuries increased rapidly with age, accounting for 2% of people aged in their 30's compared to 9% for those aged 80 years and over [38].

The road environment may contribute to the level of risk encountered by older pedestrians, as roads and pedestrian facilities are often designed based the performance of young adults [37]. According to a study in the USA, approximately 10% of older pedestrians reported that crossing roads was difficult, with the majority experiencing insufficient time to cross at signalled crossings [38]. Similar results were also found with respect to the time allowed to cross the pedestrian crossing, the availability of pedestrian crossings and the uncertainty and danger of using subways and footbridges [37]. There is also an increased likelihood that older people are taking one or more prescription medicines, which may affect their ability to use and safely interpret the road environment [37].

In Australia, pedestrian injury is a leading cause of death among 5 to 9 year old children [39]. Pedestrians aged under 10 years are particularly vulnerable because of their small physical size and underdeveloped cognitive and perceptual abilities for dealing with traffic situations [40]. A Swedish study reported that injury risk to child and adolescent pedestrians were higher among those from lower socio-economic groups [16]. The head, torso, and lower extremities are the most commonly injured body regions in vehicle-to-pedestrian crashes [41]

4.3.2. Environmental characteristics of pedestrian crashes

Road systems and motor vehicles have been designed for efficiency with little consideration for the safety of pedestrians. In the case of child pedestrians, the volume of traffic, availability of footpaths and the presence of visual obstructions have been identified as predictors of the likelihood of pedestrian injury [42]. Reduced visibility and pedestrian conspicuity may increase the incidence of crashes for pedestrians of all ages [43]. Visibility and conspicuity can be reduced by factors such as poor ambient or artificial light conditions, dark pedestrian clothing, poor weather, fog, smoke and obstructions on and near the roadway [43]. Icy weather and slippery roadways may also contribute to pedestrian injuries [14].

The most common pedestrian crashes involved a pedestrian crossing the road [44]. Locations where there are high pedestrian and traffic volumes are also associated with higher pedestrian crashes. These locations include shopping malls, on either side of a busy street and road sections close to schools and train stations [44]. A strong relationship has been found between the speed limit on the road and both number of pedestrian crashes and injury severity due to crash [36].

4.3.3. Behavioural characteristics of pedestrian crashes

Research on pedestrian behavioural factors that influence their risk of injury on the roads has been scarce. However, pedestrian alcohol consumption has been linked to the injury risk, with the presence of bars in a neighbourhood relating to an increased rate of injury to pedestrians who had been drinking [45]. Pedestrian alcohol consumption has also been associated with worse outcomes in terms of hospital and intensive care unit stay, morbidity and mortality [46].

5. Western Australian Study of Vulnerable Road User ED Presentations

5.1. Study Objectives

This study utilised hospital ED information to investigate the range of situations and events causing injury to vulnerable road users [7]. The specific objectives of the study were to:

- Describe the frequency and characteristics of motorcyclist, bicyclist and pedestrian involvement in a crash that present to an ED during a six-month period;
- Describe the injuries that occur among motorcyclist, bicyclist and pedestrian involved in a crash presenting to an ED during a six-month period;
- Determine the significance of off-road and non-motor vehicle events to the overall motorcyclist, bicyclist and pedestrian problem;
- Recommend suitable preventive behavioural and environmental strategies aimed at improving the safety for bicyclist, pedestrian and motorcyclist.

5.2. Research Design and Subjects

A six month prospective study was undertaken involving vulnerable road users who presented to the ED of the four public tertiary hospitals in Perth, Western Australia, namely, Royal Perth Hospital, Princess Margaret Hospital, Sir Charles Gardiner Hospital and Fremantle Hospital. All motorcyclists, bicyclists and pedestrians who were injured due to involvement in either a collision with a motor vehicle, a fall from a bicycle/ motorcycle or a collision with a non-motor vehicle were eligible for inclusion in the study [7].

The data collection period commenced on 1 September 2004 and was completed on 28 February 2005, as cycling and walking were prevalent during the spring and summer seasons. The following case definitions for vulnerable road users were used [14]:

- Bicyclist: Any person riding on a two or three-wheeled vehicle operated solely by pedals, but excluding mopeds, motorized scooters and other motor assisted bicycles.
- Pedestrian: Any person travelling from one location to another and injured as a result of being struck by a motor vehicle regardless of where the collision took place. The definition excludes persons using rollerblades, skateboards and other special equipment such as wheelchairs and prams.

- Motorcyclist: Any person riding a motorized two-wheeler driven by an engine with an engine capacity of 250cc and greater.

5.3. Questionnaire and Data Collection

Data were collected through a structured questionnaire. A registered nurse was responsible for recruitment and conducting all interviews. A system was set up within each of the four hospitals where the clerical staff or triage nurses contacted the research nurse whenever an injured motorcyclist, bicyclist or pedestrian presented to the ED. Individuals who had sustained a serious injury or involved in a fatal road crash were not approached. Upon receiving approval from the treating physician, a brief explanation together with an information sheet was provided to the patient. A structured interview was then scheduled after obtaining the patient's written consent. Parental consent was additionally obtained for children under the age of 16. For children aged 12 years or under, the interview was conducted in the presence of a parent who also assisted with supplying the relevant information.

The interview solicited demographic information, description of events leading to the injury, behavioural characteristics, road features and temporal factors. The questionnaire consisted of a set of "core" questions that were administered to all subjects. Each interview took about 20 minutes to complete, and demographic details were still obtained from those declined to participate.

5.4. Injury Classification and Statistical Analysis

Injuries were grouped into seven categories based on injury severity level and body region using the Abbreviated Injury Scale (AIS). The AIS is a numerical severity scoring scale which ranges from 1 (minor injury) to 6 (maximum injury - virtually unsurvivable), that also groups injuries by nine body regions. The body regions are head, spine, lower extremities, upper extremities, chest, abdomen, external, face and neck. If casualties had more than one recorded injury, the higher AIS score was used to describe the severity of injury. If casualties had more than one injury of equal severity to different body regions, the following hierarchy of body regions was used for classification: head, spine, lower extremities, chest, abdomen, upper extremities, neck, face and external. Descriptive and inferential statistics for the entire sample and for each vulnerable road user group were undertaken using the SPSS for Windows package version 14.

5.5. Sample Characteristics

Of the total 392 eligible patients, 282 consented to be interviewed, representing a response rate of 72%. No significant differences were found between participants and non-participants in terms of age ($p=0.25$) and gender ($p=0.38$). The majority of presentations came from bicyclists (n=151, 53%), followed by motorcyclists (n=104, 37%) and pedestrians (n=27, 10%). Males outnumbered females by 4 to 1 except for pedestrians (44% males).

The mean age for the entire sample was 27.5 years (SD 17.2). The mean age for males was 26.4 years (SD 17.7) and the mean age for females was 31.9 years (SD 21.9). Overall, the largest proportion of presentations was the 25 to 44 age group (n=92, 33%) followed by the 15 to 24 age group (n=68, 24%) and then the 10 to 14 age group (n=53, 19%).

Table 1 presents a breakdown of the sample by age, gender and road user type. Bicyclists (mean=23 years, SD 17.02) and motorcyclists (mean=31 years, SD 12.1) were significantly younger than pedestrians (mean=40 years, SD 24.5). Within each road user group the age distribution was also different. Half the bicyclists were less than 14 years of age (n=75), whereas 49% (n=55) of motorcyclists ranged between 25 and 44 years, and 30% (n=7) of pedestrians were 55 years or above. Moreover, a large majority of the sample were born in Australia (76%) with the remainder born in North American and European countries.

5.6. Crash characteristics

5.6.1. Crash type

Crash events were classified into the following categories:

- involvement with a motor vehicle;
- a fall/non-motor vehicle collision;
- a collision with a fixed object (e.g. tree, pole), another person or an animal.

Table 1. Sample characteristics by age group, gender and road user type

Age Group	Pedestrians		Bicyclists		Motorcyclists		
(years)	Male	Female	Male	Female	Male	Female	Total
	n (%)	n (%)	n (%)	n (%)	n (%)	n (%)	n (%)
0 to 9	0 (0)	1 (7)	19 (15)	6 (21)	0 (0)	0 (0)	26 (9)
10 to 14	2 (17)	1 (7)	43 (35)	7 (24)	0 (0)	0 (0)	53 (19)
15 to 24	4 (33)	2 (13)	23 (19)	4 (14)	30 (32)	5 (45)	68 (24)
25 to 44	3 (25)	3 (20)	24 (20)	7 (24)	51 (55)	4 (36)	92 (33)
45 to 54	2 (17)	2 (13)	7 (6)	2 (7)	10 (11)	2 (18)	25 (9)
55+	1 (8)	6 (40)	6 (5)	3 (10)	2 (2)	0 (0)	18 (6)
Total	12 (4)	15 (6)	122 (43)	29 (10)	93 (33)	11 (4)	282 (100)

Overall, only 32% (n=91) of crashes involved a collision with a motor vehicle. Fifty three percent of study participants were injured due to a fall/non-motor vehicle collision and the remainder collided with a fixed object (8%), another road user (5%) or an animal (2%). There was a significant association between the type of road user and crash type (χ^2 =61.26, p=0.01). A breakdown of crash type by road user is presented in Figure 1. All pedestrian injury events involved a motor vehicle (100%) whereas the majority of bicyclists (64%) were involved in a fall/non-motor vehicle collision. In comparison, 49% (n=51) of motorcyclists were involved in a fall/non-motor vehicle collision, 32% (n=33) a collision with a motor vehicle, and 19% (n=20) a collision with an animal (such as quokka, kangaroo, cow, dog), fixed object (tree, pole) or another road user.

5.6.2. Road type and crash location

Overall, 51% (n=144) of all crashes occurred on a public road and 49% (n=138) occurred off-road. Off-road locations refer to sidewalks, driveways, yards, cycle paths, bike trails and parking areas. There was a significant association between the type of road user and road type (χ^2=21.36, p=0.001). The majority, 58% (n=88) of bicyclist crashes, 45% (n=47) of motorcyclist crashes but only 11% (n=3) of pedestrian crashes occurred off-road. For bicyclists aged less than 14 years, 73% of crashes occurred off-road.

Public roads were further divided into highways, arterial roads and local roads, with 52% of all crashes occurring on an arterial road, 32% on a local road and 16% on a highway. In addition, 60% of all crashes on public roads occurred where there were no traffic signs or controls. The majority of crashes (88%) occurred on a public road where the speed limit was between 50 and 70 kilometres (km) per hour, 8% occurred in areas where the speed limit was >80 km and 4% occurred in areas where the speed limit was <40 km [47].

In relation to crashes occurring in off-road locations, only 10% involved a motor vehicle, 70% resulted from a fall/non-motor vehicle collision whereas 20% involved a collision with an animal/fixed object/person. The majority of off-road crashes were sustained by bicyclists (64%), followed by motorcyclists (34%) and pedestrians (2%). A large proportion occurred on footpaths (28%), followed by bush paths (15%) and in recreation areas (11%).

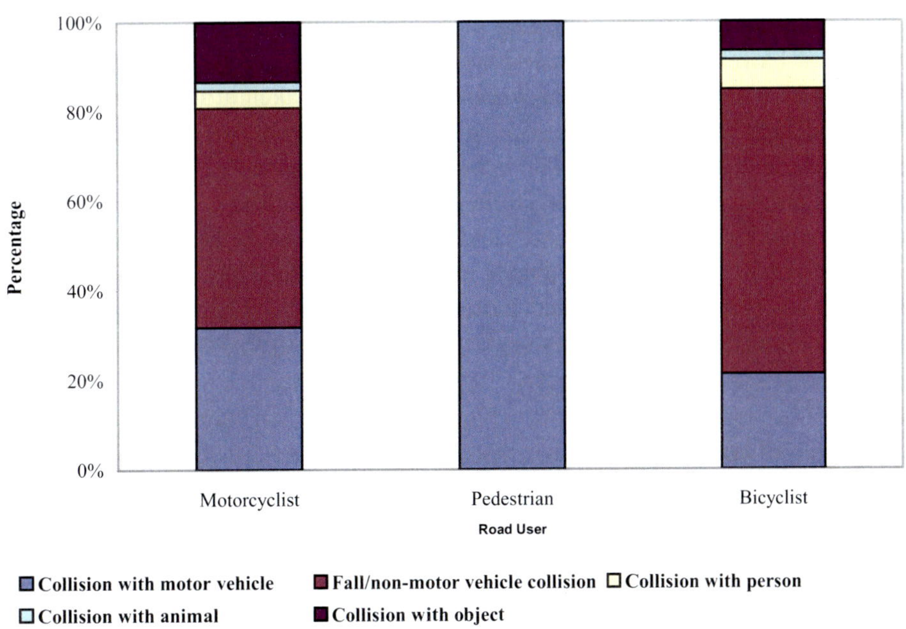

Figure 1. Crash type by road user

collision with an animal, person, object (n=22, 51%) were classified as moderate. However, collision with a motor vehicle led to the highest proportion (19%) of serious or critical injuries, compared to 16% for fall/non-motor vehicle collision and 12% for collision with an animal, person or object. The majority of injuries sustained in high speed areas were classified as moderate (46%) and serious (36%). In low (<40 kph) and medium (50 to 70 kph) speed areas, the majority of injuries were classified as minor (50% and 49% respectively) or moderate (33% and 35%).

The largest proportion of injuries reported by all vulnerable road users were to the upper extremities (37%) followed by the lower extremities (26%) and the head (18%). Injuries to the spine accounted for 5%, abdomen 3%, neck 2%, face 1% and unspecified body part, 7%. Overall, injuries to the spine accounted for all four critical injuries (100%). Serious injuries consisted mainly of injuries to the lower extremities (55%), while moderate injuries comprised mainly of injuries to the upper extremities (62%). A large proportion of minor injuries involved the head (32%).

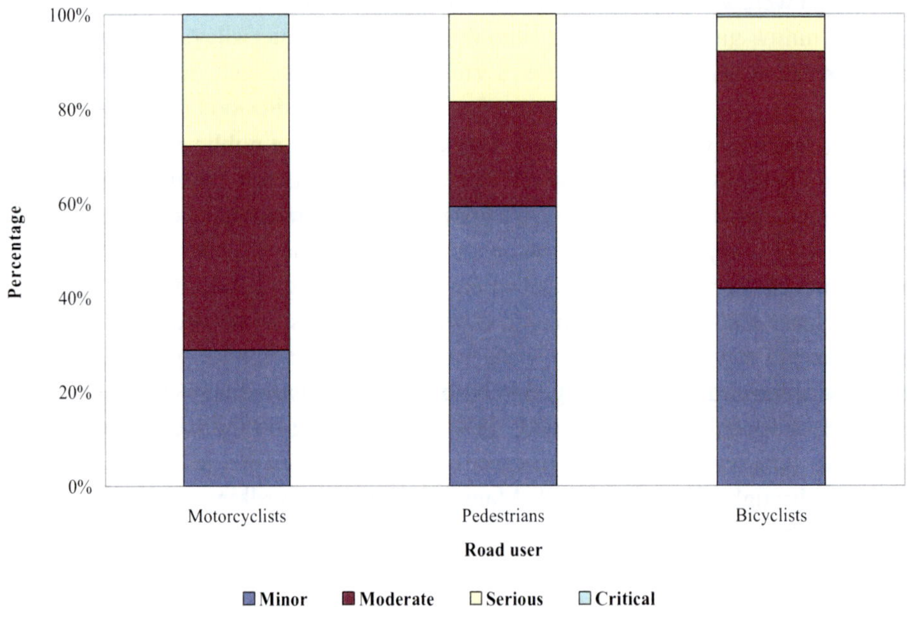

Figure 2. Severity of injury by road user

Table 2. Severity of injury by type of road

Severity of injury	Pedestrians		Bicyclists		Motorcyclists	
	Public	Off -road	Public	Off -road	Public	Off -road
Minor	14 (58)	2 (67)	30 (48)	33 (37)	21 (36)	11 (23)
Moderate	5 (21)	1 (33)	27 (42)	49 (56)	24 (43)	21 (45)
Serious	5 (21)	0 (0)	5 (8)	6 (7)	12 (21)	12 (26)
Critical	0 (0)	0 (0)	1 (2)	0 (0)	0 (0)	3 (6)
Total	24 (100)	3 (100)	63 (100)	88 (100)	57 (100)	47 (100)

5.9. Police Attendance and Reporting

Police attended only 27% of all vulnerable road user crashes of which 58% resulted in admission to hospital. Seventy percent of these police attended crashes were the result of a collision with a motor vehicle and predominantly involved motorcyclists (48%). Seventeen percent were due to a fall/non-motor vehicle collision and again, predominantly involved motorcyclists (77%). Thirteen percent involved a collision with a person/object/animal with the majority sustained by motorcyclists (60%). Bicyclists (65%) were more likely to be involved in a crash that was not reported to police compared to 31% of motorcyclists and 5% of pedestrians. Five (11%) of the injuries classified as serious/critical by attending emergency physicians were not attended by police.

5.10. Study Limitations

Several limitations should be considered when interpreting the results. These include recall error and unwillingness by the participants to admit the truth during the interview. Parents who did not witness their child's behaviour prior to the crash may also affect the validity of the responses. The potential for selection bias concerning those who sought medical care at the emergency department can lead to over representation of certain age groups. Furthermore, individual exposure data were lacking so that specific levels of risk cannot be quantified.

6. Implications on Safety Research

The Western Australian study provided valuable evidence about the characteristics of vulnerable road user crashes. It demonstrated that many bicyclists and motorcyclists who presented injured at an ED were involved in a crash that did not involve a motor vehicle and/ or occurred off-road. This finding is consistent with previous examinations of ED presentations and emphasises that these crash types pose significant injury risks [14,29,48]. It was also revealed that children under the age of 14 years were involved in a large majority of the off-road bicycle crashes.

Police attended only 27% of all vulnerable road user crashes, of which 58% were hospitalised as a result of their injuries. It is obvious that several cases in this study would not be captured in either police or hospital admission records. This lends strong support to previous literature that relying solely on police or hospital admission data will underestimate the magnitude and scope of vulnerable road user crashes and injuries [4-6,12]. Although many injuries resulting from non-motor vehicle and off-road crashes were not as severe as those involving a motor vehicle, several cases still required hospitalisation. This reinforces the need to broaden efforts to create a safe environment for vulnerable road users beyond public roadways to off-road locations. It was also found that a higher proportion of rural crashes led to severe or critical injuries than metropolitan crashes.

Behavioural factors play a pivotal role in the protection of vulnerable road users, yet information on the prevalence of behaviours associated with crash involvement or injury has been extremely limited. Despite motorcycle and bicycle helmets being compulsory in Western Australia, 32% of bicyclists and 9% of motorcyclists were not wearing them at the time of the crash. This finding is supported by past research indicating that not wearing a helmet is an injury risk factor for both motorcyclists and bicyclists [18,22,33]. Helmet use was found to be particularly low among injured adolescent bicyclists. Because of the increased risk associated with cycling and motorcycling, it is important that vulnerable road users wear adequate protection and adopt correct behaviour in traffic. Safety promotion programs targeting the waring of helmet should be aimed at both motorcyclist and bicyclists.

Vulnerable road users must make themselves clearly visible by wearing reflective items and lightly coloured clothing after dark. A substantial number of bicyclists and motorcyclists were not wearing easily visible clothing at the time of the crash, while pedestrians were involved in most of the injury events between midnight and 6 am. The effect of visibility aids such as fluorescent materials in yellow, red and orange colours are known to improve detection and recognition in the day time [49]. For night time visibility, lamps, flashing lights and retro-reflective material in red and yellow colours may also increase detection and recognition of pedestrians [50]. However, the effectiveness of these measures on drivers' response and safety of vulnerable road users has not been addressed in the literature.

Speed has been shown to affect motorcyclist crash involvement and injury severity [18,21], with 17% of motorcyclists in this study travelling at speeds above the official road limit. Also of concern, 15% of motorcyclists were unlicensed at the time of the crash. Appropriate measures to curtail speeding for motorcyclists and to increase compliance with licensing legislation are clearly needed. Furthermore, this study found that many young bicycle riders were either distracted or hurrying prior to the crash and that older pedestrians experienced perceptual difficulties when crossing roads. Training and education are essential for developing the proper attitudes, skills and road safety knowledge in these vulnerable road user groups.

7. Conclusion

Accurate information on motorcyclist, bicyclist and pedestrian crash characteristics is essential for monitoring trends as well as developing and evaluating effective road safety interventions. Knowledge of crashes involving vulnerable road users has been limited by the use of police reports which are known to be underreported, especially for non-motor vehicle and off-road collisions. Meanwhile, hospital admission data only capture the most severely injured. Therefore, information obtained from these sources does not represent the magnitude and scope of vulnerable road user crashes.

The Western Australian study provided valuable evidence on demographic, environmental and behavioural characteristics of vulnerable road users who presented at an ED as a result of their injuries. The use of a questionnaire allowed the collection of important information on behavioural factors associated with crashes that previously, was scant. Nevertheless, to improve the accuracy and representativeness of injury data, data linkage between ED presentations, hospital admissions and police records is recommended. The collection of exposure data should also be a priority in order to determine the crash and injury rates of vulnerable road users. Finally, Injury preventive measures targeting the wearing of helmet and visible clothing should be developed for both motorcyclists and bicyclists, especially among adolescent bicyclists.

ACKNOWLEDGMENTS

This research was funded by Main Roads Western Australia. The authors are grateful to Tony Radalj, Bob Peters and Brian Kidd for helpful comments, and to Claire Haworth for assistance with data collection.

REFERENCES

[1] Peden, M., Scurfield, R., Sleet, D., Mohan, D., Hyder, A., Jarawan, E. & Mathers, C. (2004). *World report on road traffic injury prevention.* Geneva: World Health Organisation.

[2] Avenoso, A. & Beckmann, J. (2005). *The Safety of Vulnerable Road Users in the Southern, Eastern and Central European Countries (The "SEC Belt").* Brussels: European Transport Safety Council.

[3] Marchant, R. J., Hill, D. L., Caccianiga, R. A. & Gant, P. D. (2008). *Reported Road Crashes in Western Australia 2006* Perth: Road Safety Council of Western Australia.

[4] Loo, B. P. & Tsui, K. L. (2007). Factors affecting the likelihood of reporting road crashes resulting in medical treatment to the police. *Inj Prev, 13(3),* 186-189.

[5] Lujic, S., Finch, C., Boufous, S., Hayen, A. & Dunsmuir, W. (2008). How comparable are road traffic crash cases in hospital admissions data and police records? An examination of data linkage rates. *Aust N Z J Public Health, 32(1),* 28-33.

[6] Dhillon, P. K., Lightstone, A. S., Peek-Asa, C. & Kraus, J. F. (2001). Assessment of hospital and police ascertainment of automobile versus childhood pedestrian and bicyclist collisions. *Accid Anal Prev, 33(4),* 529-537.

[7] Meuleners, L. B., Lee, A. H. & Haworth, C. (2006). Emergency presentations by vulnerable road users: implications for injury prevention. *Inj Prev, 12*(1), 12-14.

[8] Murray, C. J. & Lopez, A. D. (1997). Alternative projections of mortality and disability by cause 1990-2020: Global Burden of Disease Study. *Lancet, 349(9064),* 1498-1504.

[9] Beck, L. F., Dellinger, A. M. & O'Neil, M. E. (2007). Motor vehicle crash injury rates by mode of travel, United States: using exposure-based methods to quantify differences. *Am J Epidemiol, 166(2),* 212-218.

[10] Elvik, R. & Vaa, T. (2004). *The handbook of road safety measures.* Amsterdam, the Netherlands: Elsevier.

[11] Toro, K., Hubay, M., Sotonyi, P. & Keller, E. (2005). Fatal traffic injuries among pedestrians, bicyclists and motor vehicle occupants. *Forensic Sci Int, 151(2-3),* 151-156.

[12] Meuleners, L. B., Gavin, A. L., Cercarelli, L. R. & Hendrie, D. (2003). *Bicycle crashes and injuries in Western Australia 1987-2000.* Perth: Injury Research Centre, The University of Western Australia.

[13] Rosman, D. L. & Knuiman, M. W. (1994). A comparison of hospital and police road injury data. *Accid Anal Prev, 26(2),* 215-222.

[14] Stutts, J. C. & Hunter, W. W. (1999). Motor vehicle and roadway factors in pedestrian and bicyclist injuries: an examination based on emergency department data. *Accid Anal Prev, 31(5),* 505-514.

[15] Diamantopoulou, K., Brumen, I., Dyte, D. & Cameron, M. (1995). *Analysis of trends in motorcycle crashes in Victoria, Report No. 84.* Melbourne: Monash University Accident Research Centre.

[16] Hasselberg, M., Laflamme, L. & Weitoft, G. R. (2001). Socioeconomic differences in road traffic injuries during childhood and youth: a closer look at different kinds of road user. *J Epidemiol Community Health, 55(12),* 858-862.

[17] Haworth, N., Smith, R., Brumen, I. & Pronk, N. (1997). *Case-control study of motorcycle crashes.* Victoria: Monash University Accident Research Centre.

[18] Savolainen, P. & Mannering, F. (2007). Probabilistic models of motorcyclists' injury severities in single- and multi-vehicle crashes. *Accid Anal Prev, 39(5),* 955-963.

[19] Zambon, F. & Hasselberg, M. (2006). Factors affecting the severity of injuries among young motorcyclists--a Swedish nationwide cohort study. *Traffic Inj Prev, 7(2),* 143-149.

[20] Liu, B. C., Ivers, R., Norton, R., Boufous, S., Blows, S. & Lo, S. K. (2008). Helmets for preventing injury in motorcycle riders. *Cochrane Database Syst Rev*(1), CD004333.

[21] Lin, M. R., Chang, S. H., Pai, L. & Keyl, P. M. (2003). A longitudinal study of risk factors for motorcycle crashes among junior college students in Taiwan. *Accid Anal Prev, 35(2),* 243-252.

[22] Shankar, U. (2001). *Recent trends in fatal motorcycle crashes.* United States: National Center for Statistics and Analysis Research and Development.

[23] Elliott, M. A., Baughan, C. J. & Sexton, B. F. (2007). Errors and violations in relation to motorcyclists' crash risk. *Accid Anal Prev, 39(3),* 491-499.

[24] Kim, J. K., Kim, S., Ulfarsson, G. F. & Porrello, L. A. (2007). Bicyclist injury severities in bicycle-motor vehicle accidents. *Accid Anal Prev, 39(2),* 238-251.

[25] Stone, M., & Broughton, J. (2003). Getting off your bike: cycling accidents in Great Britain in 1990-1999. *Accid Anal Prev, 35(4),* 549-556.

[26] Watson, L. M. & Cameron, M. H. (2006). *Bicycle and motor vehicle crash characteristics, Report no. 251.* Melbourne: Monash University Accident Research Centre.

[27] Mehan, T. J., Gardner, R., Smith, G. A. & McKenzie, L. B. (2008). Bicycle-Related Injuries Among Children and Adolescents in the United States. *Clin Pediatr (Phila),* published online October 20, 2008, doi:2010.1177/0009922808324952.

[28] Li, G. & Baker, S. P. (1996). Exploring the male-female discrepancy in death rates from bicycling injury: the decomposition method. *Accid Anal Prev, 28(4),* 537-540.

[29] Ortega, H. W., Shields, B. J. & Smith, G. A. (2004). Bicycle-related injuries to children and parental attitudes regarding bicycle safety. *Clin Pediatr (Phila), 43(3),* 251-259.

[30] Klop, J. R. & Khattak, A. J. (1999). Factors influencing bicycle crash severity on two-lane, undivided roadways in North Carolina. *Transport. Res. Rec., 1674,* 78–85.

[31] Rodgers, G. B. (1997). Factors associated with the crash risk of adult bicyclists. *J. Saf. Res., 28(4),* 233–241.

[32] Depreitere, B., Van Lierde, C., Maene, S., Plets, C., Vander Sloten, J., Van Audekercke, R., Van der Perre, G. & Goffin, J. (2004). Bicycle-related head injury: a study of 86 cases. *Accid Anal Prev, 36(4),* 561-567.

[33] Richter, M., Otte, D., Haasper, C., Knobloch, K., Probst, C., Westhoff, J., Sommer, K. & Krettek, C. (2007). The current injury situation of bicyclists--a medical and technical crash analysis. *J Trauma, 62(5),* 1118-1122.

[34] De Lapparent, M. (2005). Individual cyclists' probability distributions of severe/fatal crashes in large French urban areas. *Accid Anal Prev, 37(6),* 1086-1092.

[35] Andersson, A. L. & Bunketorp, O. (2002). Cycling and alcohol. *Injury, 33(6)* 467-471.

[36] Garder, P. E. (2004). The impact of speed and other variables on pedestrian safety in Maine. *Accid Anal Prev, 36(4),* 533-542.

[37] Oxley, J. & Fildes, B. (2004). *Safety of older pedestrians: Strategy for Future Research and Action Initiatives, Report No. 157.* Melbourne: Monash University Accident Research Centre.

[38] Dunbar, G., Holland, C. A. & Maylor, E. A. (2004). *Older pedestrians: A critical review of the literature.* London: Department for Transportation.

[39] Al-Yaman, F., Bryant, M. & Sargeant, H. (2002). *Australia's children: their health and wellbeing 2002.* Canberra: Australian Institute of Health and Welfare.

[40] Schieber, R. A. & Vegega, M. E. (2002). Reducing childhood pedestrian injuries. *Inj Prev, 8 Suppl 1,* i1-10.

[41] Zhang, G. M., Cao, L. P., Hu, J. P. & Yang, K. H. P. (2008). A Field Data Analysis of Risk Factors Affecting the Injury Risks in Vehicle-to-Pedestrian Crashes. *Annu Proc Assoc Adv Automot Med, 52,* 199-214.

[42] Stevenson, M., Jamrozik, K. & Burton, P. (1996). A case-control study of childhood pedestrian injuries in Perth, Western Australia. *J Epidemiol Community Health, 50(3),* 280-287.

[43] Legge, M. & Rosman, D. L. (2000). *Pedestrian crashes in Western Australia, 1987-1996: The determinant of the severity of injuries.* Perth: Road Accident Prevention Research Unit.

[44] Corben, B. & Diamantopoulou, K. (1996). *Pedestrian safety issues for Victoria, Report No. 80.* Victoria: Monash University Accident Research Centre.

[45] LaScala, E. A., Gerber, D. & Gruenewald, P. J. (2000). Demographic and environmental correlates of pedestrian injury collisions: a spatial analysis. *Accid Anal Prev, 32(5),* 651-658.

[46] Small, T. J., Sheedy, J. M. & Grabs, A. J. (2006). Cost, demographics and injury profile of adult pedestrian trauma in inner Sydney. *ANZ J Surg, 76(1-2),* 43-47.

[47] Meuleners, L. B. & Haworth, C. (2005). *Motor Vehicle and Roadway Factors in Pedestrian, Bicyclist and Motorcyclist Injuries: An Examination based on Presentations to an Emergency Department in Western Australia.* Perth: Injury Research Centre, The University of Western Australia.

[48] Moll, E. K., Donoghue, A. J., Alpern, E. R., Kleppel, J., Durbin, D. R. & Winston, F. K. (2002). Child bicyclist injuries: are we obtaining enough information in the emergency department chart? *Inj Prev, 8(2),* 165-169.

[49] Kwan, I. & Mapstone, J. (2004). Visibility aids for pedestrians and cyclists: A systematic review of randomised controlled trials. *Accid Anal Prev, 36(3),* 305-312.

[50] Sayer, J. R. & Mefford, M. L. (2004). High visibility safety apparel and nighttime conspicuity of pedestrians in work zones. *J Safety Res*, *35(5),* 537-546.

In: Accidents: Causes, Analysis and Prevention
Editors: Henri Bédard and Geraud Delashmit
ISBN: 978-1-60741-712-5

Chapter 5

HUMAN FACTORS RELATED TO OCCUPATIONAL ACCIDENTS

Simo Salminen

Finnish Institute of Occupational Health, Occupational Safety, Topeliuksenkatu 41 a A, FI-00250 Helsinki, Finland

ABSTRACT

It is generally accepted that 80-90% of accidents are due to human error. Two out of three fatal occupational accidents in Australia are caused by human factor. For accident prevention, a superficial explanation about a human factor as the cause of accident is enough; we need a more detailed analysis of the nature of human factors.

The aim of this review of literature is to examine studies, where human factors related to occupational accidents were analyzed. On the one hand, well-known theories of accident causes (e.g. accident proneness) were presented with new empirical results. On the other hand, new areas of research of human factors are presented like the relationship between safety climate and occupational accidents.

Risk taking is one important area of human factors leading to occupational accidents. Risk taking theories and empirical research in the area presented. The other significant area is attributions, especially self-defensive attributions. Human errors and safety management are also meaningful topics in this review.

INTRODUCTION

The aim of this paper is to review of literature on human factors related to occupational accidents. Human factors mean here all the factors where human behavior contributes occupational safety.

THEORIES OF ACCIDENT PRONENESS AND LIABILITY

The first scientific theory attempting to explain accidents was the theory of accident proneness. Greenwood and Woods (1919) insisted that "accidents occur to limited number of individuals who have a special susceptibility to accidents" (preface). At the same time it is a first theory looking occupational accidents from human factors perspective, because it was presumed accident proneness as a stable personality trait. The theory assumed that some people have an inherent tendency to be involved in accidents (Shaw & Sichel, 1971).

There was a strong need to develop the concept of accident proneness, because it was done simultaneously in England and in Germany (Burnham, 2008). In England Eric Farmer (1925) used first time the term accident proneness, whereas in Germany Karl Marbe (1925) wrote about Unfallneigung. The authors did not know each others work until the next year.

The theory of accident proneness has been criticized on both theoretical and empirical grounds. The studies published have caused conceptual confusion concerning of the meaning of accident proneness. Accident proneness is understand as a unitary trait, as a general characteristic, an innate and unmodifiable characteristic, and is it thus used as an exploratory concept (McKenna, 1983).

The statistical criticism of the accident proneness culminates in the question of whether the entire population has the same risk to be involved in an accident or not. In practice it is very difficult to find a group of workers whose exposure to hazards is exactly the same (Mintz & Blum, 1949; Crawford, 1960; Surry, 1969). A review of 79 studies showed that accident proneness exists, but the variation in definitions of the accident proneness concept hampered the study severely (Visser et al., 2007).

Because the accident proneness theory is based on this incorrect assumption of homogeneous exposure to risk, it should be substituted by the concept of accident liability (Newbold, 1926), which is based on the assumption of the inequality of exposure (Cresswell & Froggatt, 1963). The studies of accident liability were criticized for their retrospective nature (Lawton & Parker, 1998).

To implement these theories in practice or in other words to eliminate the employees often involved in accidents did not seem to decrease the accident frequency in the company. Different tests of personality did not predict the employees who tend to be involved in accidents. In addition, eliminating the applicant by the liable to accident did not seem to decrease the accidents of the recruiting company (Guastello, 1993).

Human Factors Approach to Risk Taking Theories

The approach of human factors examined risk taking as a decision making situation. An individual actor makes the decision, does he/she take the risk or not. The theories of risk taking are examined those cognitive processes, which happen in the head of the individual actor during the risk taking situation.

Surry (1969, p. 36) has presented a decision model of the accident process, which shows how to avoid, on the one hand, hazards and, on the other hand, injuries. Both processes have six choices after which either the decision process continues or an actor falls into danger situation or an accident. The model consists of the next steps:

1. warning of danger build up (a sign of danger)
2. perception of warning
3. recognition of warning
4. recognition of avoidance mode or possibility to act in another way
5. decision to attempt to avoid
6. ability to avoid

According to Surry`s model a risk taker is a person who has gone through the first five steps in the model and chosen to approach the danger. The same steps, except the build up of warning, are the base of the risk point measure used to analyze risk taking in serious occupational accidents. The main result was that risk taking has a considerable contribution on 54% of serious occupational accidents (Salminen, 1994).

A more advanced version of Surry's model has been presented by Hale and Glendon (1987, p. 40). They divided the steps into three levels based on Rasmussen's (1986) classification: 1) Skill-based behavior represents sensorimotor performance automatically without conscious control, 2) Rule-based behavior takes place in a familiar work situation where a consciously controlled stored rule is applied, 3) Knowledge-based behavior takes place in unfamiliar situations where the goal is explicitly formulated, but the means must be found and selected according to the requirements of the situation. This classification is hierarchical, which has helped Hale and Glendon to specify the steps in their model compared to Surry's model. Secondly, the feedback loop from Hale and Hale's (1970) model makes it possible to return to the start, if the chain of steps breaks at some stage. There are nine steps in the Hale and Glendon´s model compared to the six steps of Surry's model. The most important addition was, whether an actor accepts the responsibility at the start of the action to him/herself or allocates it to others. Altogether Hale and Glendon's model can be classified as a system theoretical presentation of the phases through which a person tries to bring a danger situation into his/her control.

The previous models concern an actor's behavior in a danger situation. Contrary to these approaches, Wagenaar (1992) claimed that those who get into danger situations cannot always be said to have taken the risks, until they have fallen into these situations as a consequence of mistakes in designing and management. This view delegates risk taking from the individual level to organizational level. Empirically Wagenaar (Wagenaar, 1992; Wagenaar & Groeneweg, 1987) has analyzed accidents at sea. As a counterargument to Wagenaar, we can say that a worker using complex technology depends greatly on the boundaries of the technology and on the behavior of co-workers.

The third model of risk taking was the risk motivation theory presented by Trimpop (1994). This model assumes that both personality and situational factors influence one's risk perception, which is divided into physiological, emotional and cognitive components. Each of them has its own target level of risk, which contributes to the total target level of risk. Based on rational calculation of costs and benefits, a risk taker is motivated to act. These acts influence the environment, which then gives either positive or negative feedback. The actor can then make a new and better perception of the risks.

The Theory of Risk Compensation

The risk compensation theory assumes that people have a constant level of risk that they are ready to accept. If the risk level decreases due to safety measures (e.g. seat belts in cars), people tend to adjust their goals (e.g. try to arrive sooner at their destination) so that the risk level returns to the same level as before the measure. The effects of safety measures are thus eliminated in the long run (Peltzman, 1975; O'Neill, 1977).

The risk compensation theory has been examined mostly in traffic research, for which the seat belt laws in many countries have given a good opportunity. The results, however, have been contradictory (Jonah & Lawson, 1984; Latimer, 1992; Reinfurt et al., 1990; Rock, 1993; Adams, 1994). The discussion between Robertson (1981, 1984) and Orr (1984) is perhaps the best expression of this discrepancy, because they insisted that the same empirical data both support and reject the risk compensation. The differences in these results were caused by eliminating safety improvements in the combination of vehicle age and model year.

Some studies conducted at workplaces supported the risk compensation theory. The personal protectors used by loggers prevent injuries in those parts of the body which they protect. But at the same time, the numbers of injuries in the unprotected parts of the body increased (Klen & Väyrynen, 1984). Similarly, the introduction of power makeup equipment decreased work-related injuries significantly. At the same time, other injuries in the offshore installations increased (Mohr & Clemmer, 1989).

The problematic point in the risk compensation theory is the assumption of exact compensation (O'Neill, 1977). According to it, the driver increases his/her speed exactly as much has the safety margin given by the safety belts allows. An individual driver, however, may have any direct measure to assess the risk of a single trip. The drivers' opinion as a group about the changes in traffic risk forms over a longer time period. According to present view (Hale & Glendon, 1987), at least the extreme form of risk compensation is not true. It might be more justified to speak about partial compensation instead of exact compensation (Janssen & Tenkink, 1988). Nowadays the concept of risk compensation is often replaced by the concept of behavioural adaptation (Trimpop, 1996; Sagberg, Fosser & Saetermo, 1997).

The risk compensation theory gives an explanation for phenomena observed both in traffic safety and in occupational safety, i.e. that often a safety measure does not achieve the results have been expected. On the other hand, the theory is very simple: a human being is seen as based on the pattern of stimulus and reaction. After the introduction of a measure, people change their behavior so that the effect of the measure disappears. The idea of risk compensation has served as the origin for the development of a more complex risk homeostasis theory.

The Theory of Risk Homeostasis

According to the risk homeostasis theory, the risk of a traffic accident is determined by the homeostatic system which is analogous to the room temperature system. The main assumption of the theory is that the target level of risk is the only factor determining the accident rates. Perceived benefits and costs of risks and cautious behavior are the factors controlling the level of target risk. Risk taking can be encouraged by economic rewards or psychological benefits as for example co-workers' acceptance. Cautious behavior can be rewarded by, e.g. lower taxes (Wilde, 1982a, 1994; Wilde & Murdoch, 1982). Individuals try to reach the optimal balance between the benefits and costs of risks and cautious behavior (Wilde, 1976, 1986a).

The risk homeostasis theory argues that the effect of a safety measure depends on its motivational nature for the people who use it. The effect of a measure (e.g. mandatory seat belt use) which doesn't motivate its users will be eliminated in the long run by the drivers taking more risks (speeding or passing more dangerously). So far the risk homeostasis theory is in line with the risk compensation theory. But safety can be enhanced by measures which motivate people to desire safety (Wilde, 1982a, 1984, 1994; Simonet & Wilde, 1997). If this can be achieved, the effect of the measure may be greater than the technical calculations showed. The positive feedback from good housekeeping reduced occupational accidents at a shipyard by as much as 70-90% although only 25% of the accidents were anticipated to be preventable (Näsänen & Saari, 1987; Saari & Näsänen, 1989). However, the decreased effect of speed limits on death rates in traffic (Salusjärvi, 1980) showed that it is possible to prevent accidents also by other means than those positively motivated.

Perhaps the best field study for the risk homeostasis theory is a study conducted by Aschenbrenner, Biehl and Wurm (1989). They examined the effect of the anti-block braking systems (ABS) with taxi-drivers from Munich. The results showed that the ABS taxis were involved in slightly more accidents per time unit than other taxis. But they were seldom the originators of accidents. The risk homeostasis theory was supported by the observation that drivers with ABS drove faster and took more risks than non-ABS drivers (Aschenbrenner & Biehl, 1994; Simonet & Wilde, 1997). This study demonstrated that a purely technical safety measure without the motivation of the drivers did not improve traffic safety.

The risk homeostasis theory is based mostly on the same traffic studies as the risk compensation theory. Other researchers (Evans, 1985, 1986; McKenna, 1985a, 1985b, 1987, 1988, 1990) did not accept these studies as proof for the risk homeostasis theory. Wilde (1982b, 1984, 1985, 1986b, 1988, 1989) refuted the critique by arguing that the opposition to the risk homeostasis theory is largely due to misunderstanding of its essential propositions. He (Wilde, 1989) also said that the critics have mistakenly argued that the risk homeostasis theory denies the possibilities that education, legislation and the development of technology have on safety work. In addition, Wilde (1982b, 1989) stated that there is no empirical data questionable his theory. All in all, the discussion about the risk homeostasis theory is deadlock in a rather fruitless situation, in which both sides adhere strictly to their opinion. It has been suggested that simulation exercises with physical risk and utility might usefully be employed as a methodology to move the study of risk homeostasis theory forward (Glendon et al., 1996; Hoyes & Glendon, 1993).

The risk homeostasis theory rose out of the research on traffic accidents, but it tries to explain human behavior universally in the face of hazards. It is, in principle, possible to reapply the risk homeostasis theory to behavior in work life. The situation of a worker, however, is different from that of a driver, because a worker always acts as a part of a larger work community, whereas a driver acts more as "a free individual". The feedback given by foreman and co-workers determines the risk taking tendency of a worker. The lack of the impact of a work community is an obvious weakness in Wilde's model when considering risk taking at work (Salminen, 1991; Ruppert, 1994)

The Theory of Zero Risk

The theory of zero risk (Näätänen & Summala, 1976; Summala & Näätänen, 1988) received its name from the assumption that drivers adapt to the risks involved in driving, so that they do not generally feel any risk in a traffic situation, or their subjective risk assessment approaches zero. People avoid the feeling of risk just as they avoid pain. There is a risk threshold, above which risk is experienced as aversive. A driver feels the risk of a crash as an emotional and immediate experience, which Summala (1985, 1986, 1988) has called ostensive risk.

A driver tends to satisfy his/her motives in traffic, as well as in other spheres of life. This pushes him/her to drive faster and more hazardously. Both excitatory and inhibitory motives influence the decision making of a driver. The most hazardous excitatory motives are 'extra motives', ones outside the traffic, such as saving time and effort, which prompt the driver to increase his/her speed. The increase in the risk of an accident is related to the strength of these motives (Näätänen & Summala, 1974, 1976; Summala, 1985; Summala & Näätänen, 1988). It has been suggested that the work habits accepted by the co-workers and the need to vary actions are similar 'extra motives' in work life (Häkkinen, 1978). However, the analysis of serious occupational accidents showed that at least the strain of work community was not that kind of 'extra motive' (Salminen, 1995).

It is necessary for a theory of risk taking to have a model of decision making. According to the zero risk theory (Näätänen & Summala, 1976), the central element in a driver's decision making is a subjective risk monitor, which determines the level of accepted risk in decision making. The subjective risk level is influenced by perception, expectancy and motivation. The other central element in the model is vigilance, which must be kept sufficiently high. This is a problem especially during long, monotonous highway driving. The decision making is understood as a comparison of the strength of excitatory and inhibitory motives in one dimension (Summala & Näätänen, 1988).

The zero risk theory assumes that the tendency to take risks is based on the internal tendency of human being to maximize benefits and to do it as easily as possible. From the viewpoint of accident prevention, this means that it is very difficult to change human behavior. Instead of even attempting this, safety work ought to concentrate on eliminating hazardous situations.

Defensive Attribution Hypothesis

Attributions are our explanations for the events in everyday life (Antaki, 1982). In the past years, attributions have been one of the most often studied topics in social psychology. About 50 years after introduction of attribution theory (Heider, 1958), which has thus far remained rather divergent because the differences of opinion concerning the basic concepts (Harvey &Weary, 1984).

Unexpected and negative events, such as accidents, are particularly powerful in arousing attributions, because people need to make sense out of these events. DeJoy (1994) argued that actions to prevent accidents started more often from attributions than from actual causes.

One trend in the attribution research has been to examine the opinions about responsibility related to accidents. Walster (1966) started this research by showing that an accident victim would be assigned increasing responsibility for his/her accident as its severity increased. The literature reviews showed that the latter studies slightly confirmed the hypothesis about the attribution of responsibility (Vidmar & Crinklaw, 1974; Burger, 1981).

The research on the attribution of responsibility was given a new depth when Shaver (1970) presented the defensive attribution hypothesis. According to it, the need to protect himself/herself directs the reactions of an observer of an accident towards the victim of an accident. Personal similarity (gender, age etc.) to the accident victim lessened the observer's attribution of responsibility. The studies have mostly supported the defensive attribution hypothesis (Vidmar & Crinklaw, 1974; Burger, 1981).

One may criticize the studies of defensive attribution because they examine students' reactions to hypothetical accidents. It remains unclear whether people would react similarly in real accident situations (Bulman & Wortman, 1977). The ecological validity of the defensive attribution hypothesis requires confirmation.

A partial answer comes from a study of French telecommunication service workers' reactions to occupational accidents in which they had been victims or eyewitness (Kouabenan, 1985). The defensive attribution hypothesis is supported by the results which indicated that witnesses attribute the accident to the victims' actions, whereas the victims attribute serious accidents to bad luck. A support to the hypothesis is that the victims may also blame themselves, but only in the case of minor mishap. Later Kouabenan and his co-workers (2001) showed with employees of two French companies that foremen used more internal attributions than employees, when they explained real accidents. The self-protective attribution bias increased, when the accident severity increased. In line with defensive attribution hypothesis, the victims of serious occupational accidents in Finland attributed their accidents more often external factors than their co-workers and supervisors, whereas foremen used more often internal attributions (Salminen, 1992).

Attributions related to real accidents were also examined in Ghana. Gyekye (2003) did not find any difference in causal attributions in connection with occupational injuries between Ghanaian miners and textile workers. Gyekye and Salminen (2004), however, found support for the defensive attribution hypothesis, because Ghanaian accident victims used more often external attributions and less often internal attributions than their foremen, when they explained the occupational injuries. Co-workers of the accident victims used more often internal attributions than the victims themselves. On the other hand, co-workers explained the accident more often by external factors than the foremen. Thus, co-workers' attributions on both internal and external causal factors seem to be midway between those of the victims and their foremen.

Two Portuguese studies confirmed the relationship between attributions and real accidents. Employees of both industrial and research and development companies used more often internal attributions than external attributions. Work accident experience has a positive correlation with external causal attributions and a negative correlation with internal attributions. Attributions also mediated the relationship between accident experience and employees' behavior (Goncalves et al., 2008). Another Portuguese study showed that an accident experience was associated with external attributions supporting the defensive attributions hypothesis (Niza et al., 2008).

It is possible to examine the hypothesis of divergent perspectives by comparing victims' and foremen's opinions about the same accidents. Foremen used more often internal attributions, when the consequences of the performance were serious (Mitchell & Wood, 1980). They also evaluated the victims more responsible for the behavior leading to accident and made more internal attributions for the behavior (Mitchell & Kalb, 1981). Experienced foremen, however, tended to blame the environment more for the accidents and thus used more external attributions than inexperienced foremen (Mitchell & Kalb, 1982). The tendency of foremen to used internal attributions can also explain by the defensive attribution hypothesis, because by blaming the victim they explained their own responsibility about the accident as minimal.

The defensive attribution hypothesis is important to notice especially in accident situations where there are few hard evidences and the investigation is mostly based on the stories of eyewitnesses. The hypothesis showed that many people are more motivated to protect themselves than find out the truth in the accident situation. But based on the defensive attribution hypothesis it is possible to estimate the direction, where the statements of the eyewitnesses are distorted.

Human Error

It is generally accepted that 80-90% of accidents are due to human error (Heinrich et al., 1980; Hale & Glendon, 1987). Two out of three fatal occupational accidents in Australia were caused by human error (Williamson & Feyer, 1990; Feyer & Williamson, 1991). Although these results showed the strong contribution of Human Factors to the accidents, they are too general for accident prevention. We need a more detailed analysis of the nature of the human error.

An error means all those cases in which a planned action fails to achieve its desired outcome and this failure cannot be attributed to the intervention of some change agency (Reason & Mycielska, 1982; Reason, 1990). People have a dual role in the production system for the birth of errors on the one hand, a workers' action creates errors but on the other hand, a worker can correct his/her own mistakes of those of others and maintain normal action of the system (Hale & Glendon, 1987).

The most suitable framework for the classification of human errors at this moment is Rasmussen's (1986) SRK model:

1. Skill-based behavior represents sensorimotor performance automatically without conscious control. Work performance is based on subroutines which are subject to higher level control.
2. Rule-base behavior happens in a familiar work situation, where a consciously controlled stored rule is applied. Performance is goal-oriented, but structured by feed-forward control through a stored rule.
3. Knowledge-based behavior happens in unfamiliar situations, where the goal is explicitly formulated, based on an analysis of the environment and the overall aims of the person. The means must be found and selected according to the requirements of the situation.

The model is hierarchical, because people tend to act first at the lowest level, or skill-based behavior. If they do not succeed in the situation with this level, they next try rule-based behavior and last of all, knowledge-based behavior (Hale, 1990; Reason, 1990).

Typical errors at the skill-based level were slips from one routine to another without noticing, returning to the routine at the wrong step, and conscious control disturbing the rhythm of the action. Examples of errors in rule-based behavior were wrong classification, failure to recognize situations, and memory slips in recalling procedures. The errors in knowledge-based behavior arose, for example, from using inadequate mental models, a tendency to shift back to the routine level too soon, and focusing attention on specific aspects of the problem (Rasmussen, 1982; Reason, 1987; Hale, 1990; Wagenaar et al., 1990).

Human errors have classified according to Rasmussen's levels in four empirical studies. Out of 200 operational problems in nuclear power plants, 17% were due to skill-based error, 51% to rule-based error and 22% to knowledge-based error (Rasmussen, 1980). In Australia 55% of human errors leading to fatal occupational accident were due to skill-based behavior, 14% were due to rule-based behavior, and 14% due to knowledge-based behavior (Williamson & Feyer, 1990). Reason (1990) calculated an average of three studies on mathematical tasks, and found that there were 61% skill-based errors, 27% rule-based errors and 11% knowledge-based errors. In the analysis of Finnish fatal and serious occupational accidents skill-based errors were the most common ones, 46-66%, whereas 18-33% errors are rule-based and 5-10% knowledge-based (Salminen & Tallberg, 1996). One can conclude from these studies that the occurrence of a particular type of error depends on the nature of the task.

Based on the Rasmussen's SRK-model Reason (1990, p. 64) has developed a generic error-modelling system (GEMS). One comes inside the model from skill-based level, where the routines were checked. If there are problems, one will move to the rule-based level, where applied the stored procedures of familiar situations. If they do not be enough, the answer to the problems will be found on the knowledge-based level. Then one will examine the abstract relationships between structures and functions of action. If there is a solution, one will exit from skill-based level towards the goal. If the problem did not solve, one will start from beginning according to the feedback loop. The main assertion of the model is that people are strongly biased to search for and find a solution at the rule-based level before resorting to the far more effortful knowledge-based level.

Another, more well-known part in the Reason's (1990, p. 208) theory is so called Swiss Cheese model, which assumed that the possibility of accidents is caused by the latent failures at design and managerial levels. The consequences of these failures pierce the holes on the protective mechanisms and the accident realized. The holes on the protective mechanisms varied on their sites and sizes, which explain the relative rareness of the accidents. On the other hand, when people believed that safety procedures are in good condition, thus the holes in safety barriers increased when their actions were not checked. People were lulled into a false sense of security.

Safety Management

Safety management means all those actions made by company management and foremen to improve the safety level of the company. The focus of actions are at first hand the own workers of the company, but actions can focused also to the workers of subcontractors, clients and other interest groups. The occupational accidents are of course the most important criteria for the success of safety management, but the evaluation can also based on the assessments of employees.

The most important element in safety management is the commitment of company's top management to the occupational safety (Smith et al., 1978; Cohen & Cleveland, 1983). Even in the 10 axioms of industrial safety Heinrich (1959, p. 21) emphasized that "management has the best opportunity and ability to initiate the work of prevention..." This should be visible either as clear policy or as concrete actions; the main speeches are not enough.

The early starting point for safety management was Heinrich's thesis from year of 1931 that "the supervisor or foreman is the key man in industrial accident prevention" (Heinrich, 1959, p. 21). The characteristics of safe foremen are the same as the characteristics of successful foreman in the Finnish construction sites (Mattila, Hyttinen & Rantanen, 1994). Also the participants of serious occupational accidents stressed the contribution of foreman to improve safety (Salminen & Saari, 1995).

The third cornerstone of safety management is the participation of employees. Especially the contribution of trade union improved safety (Podgorski, 2005). On the other hand, group consultations and face-to-face discussion with employees decreased accident absenteeism (Widerszal-Bazyl & Warszewska-Makuch, 2008). Higher level of employee empowerment improved employees' safety awareness and reporting safety problems (Hsu et al., 2008).

Safety Climate

Safety climate is one way to measure the effectiveness of safety management. If the safety management has been successful, then the employees assessed the safety climate of the company positively. In the company with good safety climate occurred less accidents than in the company with poor safety climate (Donald & Canter, 1994; Mearns, Whitaker & Flin, 2003; Felknor et al., 2000; Varonen & Mattila, 2000).

There is no consensus about the definition of safety climate among researchers. That is why nine different definitions of safety climate with key words like group, employees' perceptions, assembly, and molar were found (Guldenmund, 2000). One way is to define safety climate as the employees' perceptions, attitudes, and beliefs about the attitudes of the company's management towards risks and safety (Salminen & Seppälä, 2005). Safety culture is a more complex trait reflecting fundamental values, norms and expectations in the company (Mearns & Flin, 1999).

The scientific measurement of safety climate was started by Zohar (1980), when he developed a questionnaire with 40 items in Israel. In the factor analysis there were eight factors, of which two most important ones were perceived relevance of safety to work behavior, and the employees' perceptions about the management's attitudes towards safety. Later studies have only partially confirmed the factor structure found by Zohar, because according to a review of 18 studies management, safety systems and risk were the most often found factors (Flin, Mearns, O'Connor & Bryden, 2000).

The relationship between safety climate and occupational accidents has been examined based on the assumption that there are fewer accidents in the companies with good safety climate than those with poor safety climate. The assumption has confirmed by 16 studies done in eight countries (Salminen & Seppälä, 2005). Only in the study among Australian road construction workers no connection was found between safety climate and occupational accidents (Glendon & Litherland, 2001).

In Finland safety climate was measured with a Finnish questionnaire developed by Seppälä (1992), and based on Zohar's original questionnaire. When the safety attitudes of employees in plywood factories, shipyards, and forestry and construction sites were measured, the attitudes loaded on four factors: organizational responsibility, workers' concern about safety, workers' indifference in regard to safety, and the level of safety actions. Measurement of safety climate in road administration with the same questionnaire resulted in four factors: attitudes towards safety in organization, changes in work demands, appreciation of the work, and safety as a part of productive work (Niskanen, 1994). Safety climate evaluated by wood-processing workers was also loaded on four factors: organizational responsibility, workers' safety attitudes, supervision of safety, and the company's safety precautions (Varonen & Mattila, 2000). In the comparison of Finnish- and Swedish-speaking workers in Finland, safety management and personal and group-level safety behavior were the most heavily loaded two factors for both language groups (Salminen & Seppälä, 2005). These Finnish studies with the same questionnaire showed that although the number of factories was the same, the content of the factories varied. This result indicated that the safety climate measure and the factor analysis based on it are very context dependent.

IMPLEMENTATION OF HUMAN FACTORS

Accident prevention happens mainly at the shop floor level. Thus the implementation of human factors to the work life happens mostly at the same level. This paper gives only theoretical background for this work.

One problem of human factors researchers is to change their research findings to financial values. Decision makers in the companies ask about the economical benefits of human factors investigations to the company (Kerr et al., 2008). In most cases, however, it is difficult to calculate the financial value of human factor.

REFERENCES

Adams, J. G. U. (1994). Seat belt legislation: the evidence revisited. *Safety Science, 18,* 135-152.

Antaki, C. (1982). A brief introduction to attribution and attributional theories. In C. Antaki & C. Brewin (Eds.), *Attributions and psychological change* (pp. 3-21). London: Academic Press.

Aschenbrenner, M., Biehl, B. & Wurm, G. (1989). Einfluss der Risikokompensation auf die Wirkungen von Verkehrssicherheitsmassnahmen am Beispiel ABS. In B. Ludborzs (Ed.), *Psychologie der Arbeitssicherheit* (pp. 150-160). Heidelberg, Germany : Asanger.

Aschenbrenner, K. M. & Biehl, B. (1994). Improved safety through improved technical measures? In R. M. Trimpop, & G. J. S. Wilde (Eds.), *Challenges to accident prevention* (pp. 81-89). Groningen, Netherlands: Styx Publications.

Bulman, R. J. & Wortman, C. B. (1977). Attributions of blame and coping in the "Real World":

Severe accident victims react to their lot. *Journal of Personality and Social Psychology, 35,* 351-363.

Burger, J. M. (1981). Motivational biases in the attribution of responsibility for an accident: a meta-analysis of the defensive-attribution hypothesis. *Psychological Bulletin, 90,* 496-512.

Burnham, J. C. (2008). Accident proneness (Unfallneigung): A classical case of simultaneous discovery/construction in psychology. *Science in Context, 21,* 99-118.

Cohen, H. H. & Cleveland, R. J. (1983). Safety program practices in record-holding plants. *Professional Safety, 28(3),* 26-33.

Crawford, P. L. (1960). Hazard exposure differentiation necessary for the identification of the accident-prone employee. *Journal of Applied Psychology, 44,* 192-194.

Cresswell, W. L. & Froggatt, P. (1963). *The causation of bus driver accidents. An epidemiological study.* London: Oxford University Press.

DeJoy, D. M. (1994). Managing safety in the workplace: an attribution theory analysis and model. *Journal of Safety Research, 25,* 3-17.

Donald, I. & Canter, D. (1994). Employee attitudes and safety in the chemical industry. *Journal of Loss Prevention in Process Industry, 7,* 203-208.

Evans, L. (1985). Human behavior feedback and traffic safety. *Human Factors, 27,* 555-576.

Evans, L. (1986). Risk homeostasis theory and traffic accident data. *Risk Analysis, 6,* 81-94.

Farmer, E. (1925) *The method of grouping by differential tests in relation to accident proneness.* Industrial Fatigue Research Board, Annual report. p. 43-45.

Felknor, S. A., Aday, L. A., Burau, K. D., Delclos, G. L. & Kapadia, A. S. (2000). Safety climate and its association with injuries and safety practices in public hospitals in Costa Rica. *International Journal of Occupational and Environmental Health, 6,* 18-25.

Feyer, A.-M. & Williamson, A. M. (1991). A classification system for causes of occupational accidents for use in preventive strategies. *Scandinavian Journal of Work, Environment and Health, 17,* 302-311.

Flin, R., Mearns, K., O'Connor, P. & Bryden, R. (2000). Measuring safety climate: identifying the common features. *Safety Science, 34,* 177-192.

Glendon, A. I., Hoyes, T. W., Haigney, D. E. & Taylor, R. G. (1996). A review of risk homeostasis theory in simulated environments. *Safety Science, 22,* 15-25.

Glendon, A. I. & Litherland, D. K. (2001). Safety climate factors, group differences and safety behaviour in road construction. *Safety Science, 39,* 157-188.

Goncalves, S. M. P., da Silva, S. A., Lima, M. L. & Meliá, J. L. (2008). The impact of work accidents experience on causal attributions and worker behaviour. *Safety Science, 46,* 992-1001.

Greenwood, M. & Woods, H.M. (1919). *The incidence of industrial accidents upon individuals with special reference to multiple accidents*. Report No. 4, Medical Research Committee, Industrial Fatigue Research Board. London: Her Majesty's Stationery Office.

Guastello, S. J. (1993). Do we really know how well our occupational accident prevention programs work? *Safety Science, 16,* 445-463.

Guldenmund, F. W. (2000). The nature of safety culture: a review of theory and research. *Safety Science, 34,* 215-257.

Gyekye, S. A. (2003). Causal attributions of Ghanaian industrial workers for accident occurrence: Miners and non-miners perspective. *Journal of Safety Research, 34,* 533-538.

Gyekye, S. A. & Salminen, S. (2004). Causal attributions of Ghanaian industrial workers for accident occurrence. *Journal of Applied Social Psychology, 34,* 2324-2342.

Hale, A. R. (1990). Safety rules o.k.? Possibilities and limitations in behavioural safety strategies. *Journal of Occupational Accidents, 12,* 3-20.

Hale, A. R. & Glendon, A. I. (1987). *Individual behaviour in the control of danger.* Amsterdam: Elsevier.

Hale, A. R. & Hale, M. (1970). Accidents in perspective. *Occupational Psychology, 44,* 115-121.

Harvey, J. H. & Weary, G. (1984). Current issues in attribution theory and research. *Annual Review of Psychology, 35,* 427-459.

Heider, F. (1958). *The psychology of interpersonal relations.* New York: Wiley.

Heinrich, H. W. (1959). *Industrial accident prevention.* New York: McGraw-Hill.

Heinrich, H. W., Petersen, D. & Roos, N. (1980). *Industrial accident prevention (5th ed.).* New York: McGraw-Hill.

Hoyes, T. W. & Glendon, A. I. (1993). Risk homeostasis: issues for future research. *Safety Science, 16,* 19-33.

Hsu, S. H., Lee, C.-C., Wu, M.-C. & Takano, K. (2008). A cross-cultural study of organizational factors on safety: Japanese vs. Taiwanese oil refinery plants. *Accident Analysis and Prevention, 40,* 24-34.

Häkkinen, S. (1978). *Tapaturmateoriat ja niiden kehittäminen [Accident theories and developing of them].* Otaniemi: Helsinki University of Technology, Laboratories of Industrial Economics and Industrial Psychology, Report No. 36. [in Finnish]

Janssen, W. & Tenkink, E. (1988). Risk homeostasis theory and its critics: time for an agreement. *Ergonomics, 31,* 429-433.

Jonah, B. A. & Lawson, J. J. (1984). The effectiveness of the Canadian mandatory seat belt use laws. *Accident Analysis and Prevention, 16,* 433-450.

Kerr, M. P., Knott, D. S., Moss, M. A., Clegg, C. W. & Horton, R. P. (2008). Assessing the value of human factors initiatives. *Applied Ergonomics, 39,* 305-315.

Klen, T. & Väyrynen, S. (1984). The role of personal protection in the prevention of accidental injuries in logging work. *Journal of Occupational Accidents, 6,* 263-275.

Kouabenan, D. R. (1985). Degree of involvement in an accident and causal attribution. *Journal of Occupational Accidents, 7,* 187-194.

Kouabenan, D. R., Gilibert, D., Médina, M. & Bouzon, F. (2001). Hierarchical position, gender, accident severity, and causal attribution. *Journal of Applied Social Psychology, 31,* 553-575.

Latimer, E. A. (1992). Effects of increased auto safety belt use levels on fatalities. *Risk Analysis, 12,* 449-454.

Lawton, R. & Parker, D. (1998). Individual differences in accident liability: A review and integrative approach. *Human Factors, 40,* 655-671.

Marbe, K. (1925). Zur praktischen Psychologie der Unfälle und Betriebsunglücke. *Verhandlungen der phys.-med. Gesellschaft zu Würzburg,* 172-175.

Mattila, M., Hyttinen, M. & Rantanen, E. (1994). Effective supervisory behaviour and safety at the building site. *International Journal of Industrial Ergonomics, 13,* 85-93.

McKenna, F. P. (1983). Accident proneness: a conceptual analysis. *Accident Analysis and Prevention, 15,* 65-71.

McKenna, F. P. (1985a). Do safety measures really work? An examination of risk homoeostasis theory. *Ergonomics, 28,* 489-498.

McKenna, F. P. (1985b). Evidence and assumptions relevant to risk homoeostasis. *Ergonomics, 28,* 1539-1541.

McKenna, F. P. (1987). Behavioural compensation and safety. *Journal of Occupational Accidents, 9,* 107-121.

McKenna, F. P. (1988). What role should the concept of risk play in theories of accident involvement? *Ergonomics, 31,* 469-484.

McKenna, F. P. (1990). In defence of conventional safety measures: A reply to G.J.S. Wilde. *Journal of Occupational Accidents, 11,* 171-181.

Mearns, K. J. & Flin, R. (1999). Assessing the state of organizational safety - culture or climate? *Current Psychology, 18,* 5-18.

Mearns, K., Whitaker, S. M. & Flin, R. (2003). Safety climate, safety management practice and safety performance in offshore environments. *Safety Science, 41,* 641-680.

Mintz, A. & Blum, M. L. (1949). A re-examination of the accident proneness concept. *Journal of Applied Psychology, 33,* 195-211.

Mitchell, T. R. & Kalb, L. S. (1981). Effects of outcome knowledge and outcome valence on supervisors` evaluations. *Journal of Applied Psychology, 66,* 604-612.

Mitchell, T. R. & Kalb, L. S. (1982). Effects of job experience on supervisor attributions for a subordinate`s poor performance. *Journal of Applied Psychology, 67,* 181-188.

Mitchell, T. R. & Wood, R. E. (1980). Supervisor`s responses to subordinate poor performance: a test of an attributional model. *Organizational Behavior and Human Performance, 25,* 123-138.

Mohr, D. L. & Clemmer, D. I. (1989). Evaluation of an occupational injury intervention in the petroleum drilling industry. *Accident Analysis and Prevention, 21,* 263-271.

Newbold, E. M. (1926). A contribution to the study of the human factor in the causation of accidents. Reproduced in W. Haddon, E. A. Suchman, & D. Klein (Eds.), *Accident research* (pp. 397-410). New York: Harper & Row, 1964.

Niskanen, T. (1994). Safety climate in the road administration. *Safety Science, 17,* 237-255.

Niza, C., Silva, S. & Lima, M. L. (2008). Occupational accident experience: Association with workers' accident explanation and definition. *Safety Science, 46,* 959-971.

Näsänen, M. & Saari, J. (1987). The effects of positive feedback on housekeeping and accidents at a shipyard. *Journal of Occupational Accidents, 8,* 237-250.

Näätänen, R. & Summala, H. (1974). A model for the role of motivational factors in drivers' decision-making. *Accident Analysis and Prevention, 6,* 243-261.

Näätänen, R. & Summala, H. (1976). *Road-user behavior and traffic accidents.* Amsterdam: North-Holland.

O'Neill, B. (1977). A decision-theory model of danger compensation. *Accident Analysis and Prevention, 9,* 157-165.

Orr, L. D. (1984). The effectiveness of automobile safety regulation: evidence from the FARS data. *American Journal of Public Health, 74,* 1384-1389.

Peltzman, S. (1975). The effects of automobile safety regulation. *Journal of Political Economy, 83,* 677-725.

Podgorski, D. (2005). Workers' involvement - A missing component in the implementation of occupational safety and health management systems in enterprises. *International Journal of Occupational Safety and Ergonomics, 11,* 219-231.

Rasmussen, J. (1980). What can be learned from human error reports? In K. D. Duncan, M. M. Gruneberg, & D. Wallis (Eds.), *Changes in working life* (pp. 97-113). Chichester: Wiley.

Rasmussen, J. (1982). Human errors. A taxonomy for describing human malfunction in industrial installations. *Journal of Occupational Accidents, 4,* 311-333.

Rasmussen, J. (1986). *Information processing and human-machine interaction.* New York: North-Holland.

Reason, J. (1987). Generic error-modelling system (GEMS): A cognitive framework for locating common human error forms. In J. Rasmussen, K. Duncan, & J. Leplat (Eds.), *New technology and human error* (pp. 63-83). Chichester: Wiley.

Reason, J. (1990). *Human error.* Cambridge: Cambridge University Press.

Reason, J. & Mycielska, K. (1982). *Absent-minded? The psychology of mental lapses and everyday errors.* Englewood Cliffs, NJ: Prentice-Hall.

Reinfurt, D. W., Campbell, B. J., Stewart, J. R. & Stutts, J. C. (1990). Evaluating the North Carolina safety belt wearing law. *Accident Analysis and Prevention, 22,* 197-210.

Robertson, L. S. (1981). Automobile safety regulations and death reductions in the United States. *American Journal of Public Health, 71,* 818-822.

Robertson, L. S. (1984). Automobile safety regulation: rebuttal and new data. *American Journal of Public Health, 74,* 1390-1394.

Rock, S. M. (1993). Risk compensation and the Illinois seat belt use law. *Accident Analysis and Prevention, 25,* 537-544.

Ruppert, F. (1994). Risk-compensation - a theoretical approach suitable to occupational health and safety? In R. M. Trimpop, & G. J. S. Wilde (Eds.), *Challenges to accident prevention* (pp. 109-118). Groningen, The Netherlands: Styx Publications.

Saari, J. & Näsänen, M. (1989). The effect of positive feedback on industrial housekeeping and accidents; a long-term study at a shipyard. *International Journal of Industrial Ergonomics, 4,* 201-211.

Sagberg, F., Fosser, S. & Saetermo, I.-A. F. (1997). An investigation of behavioural adaptation to airbags and antilock brakes among taxi drivers. *Accident Analysis and Prevention, 29,* 293-302.

Salminen, S. (1991). Risk homeostasis: applications in working life - for and against. In M. Salusjärvi (Ed.), *Riskhomeostasis*? (pp. 31-38). VTT symposium **124**. Espoo, Finland: Valtion teknillinen tutkimuskeskus.

Salminen, S. (1992). Defensive attribution hypothesis and serious occupational accidents. *Psychological Reports, 70,* 1195-1199.

Salminen, S. (1994). Risk taking and serious occupational accidents. *Journal of Occupational Health and Safety - Australia and New Zealand, 10,* 267-274.

Salminen, S. (1995). Does pressure from the work community increase risk taking? *Psychological Reports, 77,* 1247-1250.

Salminen, S. & Saari, J. (1995). Measures to improve safety and productivity simultaneously. *International Journal of Industrial Ergonomics, 15,* 261-269.

Salminen, S. & Seppälä, A. (2005). Safety climate in Finnish- and Swedish-speaking companies. *International Journal of Occupational Safety and Ergonomics, 11,* 389-397.

Salminen, S. & Tallberg, T. (1996). Human errors in fatal and serious occupational accidents in Finland. *Ergonomics, 39,* 980-988.

Salusjärvi, M. (1980). The experiments with recommended and compulsory speed limits in Finland during 1969-1976. *International Review of Applied Psychology, 29,* 517-528.

Seppälä, A. (1992). *Turvallisuustoiminta, sen kehittäminen ja yhteydet työtapaturmiin [Evaluation of safety measures, their improvement and connections to occupational accidents].* Työ ja ihminen, Supplement 1/92. Helsinki, Finland: Työterveyslaitos.

Shaver, K. G. (1970). Defensive attribution: effects of severity and relevance on the responsibility assigned for an accident. *Journal of Personality and Social Psychology, 14,* 101-113.

Shaw, L. & Sichel, H. S. (1971). *Accident proneness.* Oxford: Pergamon Press.

Simonet, S. & Wilde, G. J. S. (1997). Risk: Perception, acceptance and homeostasis. *Applied Psychology: An International Review, 46,* 235-252.

Smith, M. J., Cohen, H. H., Cohen, A. & Cleveland, R. J. (1978). Characteristics of successful safety programs. *Journal of Safety Research, 10,* 5-15.

Summala, H. (1985). Modeling driver behavior: a pessimistic prediction? In L. Evans, & R. C. Schwing (Eds.), *Human behavior and traffic safety* (pp. 43-61). New York: Plenum.

Summala, H. (1986). The deterministic man in a stochastic world: risk management on the road. In B. Brehmer, H. Jungermann, P. Lourens, & G. Sevón (Eds.), *New directions in research on decision making* (pp. 275-285). Amsterdam: Elsevier.

Summala, H. (1988). Risk control is not risk adjustment: the zero-risk theory of driver behaviour and its implications. *Ergonomics, 31,* 491-506.

Summala, H. & Näätänen, R. (1988). The zero-risk theory and overtaking decisions. In T. Rothengatter, & R. de Bruin (Eds.), *Road user behaviour: theory and research* (pp. 82-92). Assen/Maastricht, The Netherlands: Van Gorcum.

Surry, J. (1969). *Industrial accident research: A human engineering appraisal.* Toronto: University of Toronto, Department of Industrial Engineering.

Trimpop, R. M. (1994). *The psychology of risk taking behavior.* Amsterdam: North-Holland.

Trimpop, R. M. (1996). Risk homeostasis theory: Problems of the past and promises for the future. *Safety Science, 22,* 119-130.

Varonen, U. & Mattila, M. (2000). The safety climate and its relationship to safety practices, safety of the work environment and occupational accidents in eight wood-processing companies. *Accident Analysis and Prevention, 32,* 761-769.

Vidmar, N. & Crinklaw, L. D. (1974). Attributing responsibility for an accident: a methodological and conceptual critique. *Canadian Journal of Behavioral Sciences, 6,* 112-130.

Visser, E., Pijl, Y. J., Stolk, R. P., Neeleman, J. & Rosmalen, J. G. M. (2007). Accident proneness, does it exist? A review and meta-analysis. *Accident Analysis and Prevention, 39,* 556-564.

Wagenaar, W. A. (1992). Risk taking and accident causation. In: J. F. Yates (Ed.), *Risk-taking behavior* (pp. 257-281). Chichester: Wiley.

Wagenaar, W. A., Hudson, P. T. W. & Reason, J. T. (1990). Cognitive failures and accidents. *Applied Cognitive Psychology, 4,* 273-294.

Wagenaar, W. A. & Groeneweg, J. (1987). Accidents at sea: Multiple causes and impossible consequences. *International Journal of Man-Machine Studies, 27,* 587-598.

Walster, E. (1966). Assignment of responsibility for an accident. *Journal of Personality and Social Psychology, 3,* 73-79.

Widerszal-Bazyl, M. & Warszewska-Makuch, M. (2008). Employee direct participation in organisational decisions and workplace safety. *International Journal of Occupational Safety and Ergonomics, 14,* 367-378.

Wilde, G. J. S. (1976). Social interaction patterns in driver behavior: an introductory review. *Human Factors, 18,* 477-492.

Wilde, G. J. S. (1982a). The theory of risk homeostasis: implications for safety and health. *Risk Analysis, 2,* 209-225.

Wilde, G. J. S. (1982b) Critical issues in risk homeostasis theory. *Risk Analysis, 2,* 249-258.

Wilde, G. J. S. (1984). Evidence refuting the theory of risk homoeostasis? A rejoinder to Frank P. McKenna. *Ergonomics, 27,* 297-304.

Wilde, G. J. S. (1985). Assumptions necessary and unnecessary to risk homoeostasis. *Ergonomics, 28,* 1531-1538.

Wilde, G. J. S. (1986a) Beyond the concept of risk homeostasis: suggestions for research and application towards the prevention of accidents and lifestyle-related disease. *Accidents Analysis and Prevention, 18,* 377-401.

Wilde, G. J. S. (1986b). Notes on the interpretation of traffic accident data and of risk homeostasis theory: a reply to L. Evans. *Risk Analysis, 6,* 95-101.

Wilde, G. J. S. (1988). Risk homeostasis theory and traffic accidents: propositions, deductions and discussion of dissension in recent reactions. *Ergonomics, 31,* 441-468.

Wilde, G. J. S. (1989). Accident countermeasures and behavioural compensation: the position of risk homeostasis theory. *Journal of Occupational Accidents, 10,* 267-292.

Wilde, G. J. S. (1994). *Target risk.* Toronto: PDE Publications.

Wilde, G. J. S. & Murdoch, P. A. (1982). Incentive systems for accident-free and violation-free driving in the general population. *Ergonomics, 25,* 879-890.

Williamson, A. & Feyer, A. M. (1990). Behavioural epidemiology as a tool for accident research. *Journal of Occupational Accidents, 12,* 207-222.

Zohar, D. (1980). Safety climate in industrial organizations: Theoretical and applied implications. *Journal of Applied Psychology, 65,* 96-102.

In: Accidents: Causes, Analysis and Prevention
Editors: Henri Bédard and Geraud Delashmit
ISBN: 978-1-60741-712-5

Chapter 6

SLEEP RELATED CAR CRASHES: RISK PERCEPTION AND COPING WITH SLEEPINESS AT THE WHEEL

Fabio Lucidi, Luca Mallia and Cristiano Violani
Department of Psychology, "Sapienza" University of Rome, Rome, Italy

ABSTRACT

Sleepiness is often a contributing factor in car crashes. Despite drivers under 30 years old are particularly prone to sleep-related accidents in the early morning, they often underestimate the role of sleepiness as a risk factor for car crashes. The aim of the study was to examine which factors are associated with the risk perception to have night-time car crashes in young drivers and to evaluate the strategies most commonly used to counteract sleepiness at the wheels. 1,123 young drivers (18-25 years old) with at least a 6-months driving experience participated to the study. Results indicated that males are less worried about night-time car crashes than females. The perceived likelihood of having a sleep-related car crash is low for those who report driving at night frequently. On the other hand, the level of concern regarding sleep-related car crashes increases in young people who had experienced sleepiness-impaired driving.

Keywords: Night-time Driving, Risk Perception, Sleepiness at the wheels, Young Drivers.

INTRODUCTION

Sleepiness is often a contributing factor to car crashes, but the lack of a set of standard criteria for estimating the human and financial costs of sleep-related car accidents tends to create disagreement among research studies on this topic (Zulley et al., 1994; Horne & Reyner, 1995a; Leger, 1995; Pack et al., 1995; Webb, 1995; Lyznicki, 1998). Connor and colleagues (2002) suggested that a significantly increased risk is associated with 1) drivers who identified themselves as sleepy (Stanford sleepiness score 4–7 versus 1–3), 2) drivers who reported 5 hours or less of sleep in the previous 24 hours, and 3) driving between 2 a.m. and 5 a.m. compared to other times of day. The relevance of the last aspect is emphasized in several studies (Prokop & Prokop, 1955; Langolis et al., 1986; Mitler et al., 1988; Zomer & Lavie, 1990; Summala & Mikkola; 1994; Pack et al., 1995; Folkard, 1997). Drivers under 30 years of age (especially men) are particularly prone to sleep-related car accidents in the early morning (Horne & Reyner, 1995b; Lyznicki, 1998; McConnel, 2003), even when alcohol is ruled out as a causal factor (Akerstedt et al., 1994). Akerstedt and Kecklund (2001) indicated that young drivers (18–24 years old) were 5–10 times more likely to be involved in an accident late at night compared to forenoon. The main reason why there is a high incidence of crashes involving young people during the early hours of the morning is that these drivers are the most prevalent road users at this time of the day (Horne & Reyner, 1995b). Furthermore, data are available indicating that drivers under 30 years of age accumulate a great sleep debt before long holiday trips (Philip et al., 1996). It is also reasonable to assume that drivers in this age range have less experience and knowledge about how to cope with fatigue while driving (Summala & Mikkola, 1994).

Despite the growing body of evidence indicating that young drivers are a population at-risk for nocturnal car accidents caused by sleepiness, they usually tend to consider sleepiness a rather trivial matter instead of a serious risk in terms of safety (Dinges, 1990). Carskadon (2002) interviewed drivers between 16 and 20 years old and found that a large number of them (67% of the sample) reported to have driven while impaired by sleepiness. Several studies have also reported that young novice drivers (18–21 years old) are characterized by perceiving relatively low levels of risk in specific driving situations compared to other groups of drivers (e.g. Sivak et al, 1989; Deery, 1999). Young drivers (18–24 years old) often underestimate potential hazards in the environment and overestimate their driving ability (Gregersen, 1996). The issue of risk perception in novice drivers was addressed by several studies (for a review see Rothengatter, 2002). Just few studies about risk perception included women drivers (e.g. Harré, 2000). DeJoy (1992) found that young males (18–24 years old) considered risky behavior lesser likely to lead to an accident than did young females. Regardless of gender, different models (Summala, 1997; Fuller, 2000; Kuiken & Twisk, 2001) and studies (Groeger, 2000) suggest that the overestimation of driving ability and the underestimation of the risk of car crashes in young drivers increase with the increase of driving experience. Unfortunately, as far as we know, no systematic study has analyzed whether gender and past driving experience are related to the specific risk perception of sleep related crashes in young drivers.

Some studies have considered the coping strategies used to counteract sleepiness at the wheels: young drivers usually believe that they can remain alert during times of the day in which the endogenous pressure to sleep might be high by means of a compensatory effort, even if the data indicate that this strategy is not effective in counteracting periods of poor, inefficient and variable performance (Johnson, 1982; Dinges & Kribb, 1991; Dinges, 1992). According to Jennings et al. (2003), sleep pressure impairs attentive processes that guide and control actions, and in a state of sleep deprivation the effort to cope with sleepiness do not appear to be expended readily. No systematic study has simultaneously analyzed the different strategies adopted by young drivers to counteract sleepiness at the wheels and participants' past experience related to night-time driving.

The aim of the present study was to examine which main factors are associated with the risk perception to have night-time car crash in young drivers and to evaluate the strategies most commonly used to counteract sleepiness at the wheels.

METHOD

Participants

Participants were recruited among first, second and third year undergraduate students from different programs at the University of Rome "La Sapienza". Students were also asked to name acquaintances who could be willing in participating. Participants were invited to take part to a study aimed at evaluating their beliefs regarding the effects of sleepiness on nighttime road accidents. 1,123 young drivers (58.2% female ; 41.8 % males), between the ages of 18 and 25 years (mean=21.04, SD=1.65), participated to the study. All of them had a driver's license and at least 6 months of driving experience (mean =36.34 months; SD= 23.8 months)

Procedure

Trained research assistants, after contacting the students and obtaining their consent to participate to the study, interviewed them individually to administer a questionnaire addressing the topics of interest to the study. The interview lasted about 20 minutes.

Instruments

The questionnaire was composed of two different sections concerning respectively:

1) *Questions about night-time driving and the related risk perceptions.* Participants were asked to report how many times in the last 6 months they had driven between midnight and 5a.m. ("Never", "Up to 2 times a month", "2-4 times a month", "More than 4 times a month"). Participants were also asked to evaluate the likelihood of being involved in a night-time car accident due to sleepiness and to indicate their level of concern about this possibility. Rating scales from 1 to 10 (from "very low" to "very high") were used for both questions.
2) *Questions about previous experience of sleepiness at the wheels and about the strategies used to cope with it.* The participants were asked to indicate whether they had experienced sleepiness during driving in the last 6 months ("Yes", "No"). Respondents who answered "Yes", were asked to indicate how many sleepiness episodes they had experienced. Finally, participants were asked to imagine a situation in which, while driving at night, they suddenly felt that sleepiness would affect their driving. They were then asked to choose what they would have done in such a situation, considering one of three options: a) "Ignore the problem and continue driving"; b) "Continue driving while doing something to cope with the problem"; c) "Stop driving and do something to cope with the problem". When the participants chose either the second or the third option, they were also asked to select, from a list provided, one or more appropriate courses of action they would adopt.

Statistical Analyses

The Chi^2 test was computed in order to analyze across-gender differences on the frequency of night-time driving in the last 6 months.

A linear regression was performed to examine which factors influence the perceived risk of a car accident due to sleepiness considering gender, frequency of night-time driving and past experience of sleepiness during driving as independent variables. The same predictors were also used in a logistic regression modeling their effects on the coping strategies adopted to counteract sleepiness at the wheels.

RESULTS

As illustrated in Table 1, there was a statistically significant difference in the frequency of night-time driving between male and female young drivers ($Chi^2_{(3)}$= 80.35; $p<0.001$). In particular, male drivers reported to drive significantly more often than female drivers between midnight and 5a.m.

As illustrated in Table **2**, males described themselves as less worried about night-time car crashes than did females (β=.07). Such concerns decreased with the increasing of night-time driving (β=-.12), whereas the frequency of episodes of sleepiness during driving increased the perception of this risk (β=.14).

Regarding the coping strategies used to counteract sudden sleepiness at the wheels, nearly 52% of the participants answered that they "would continue driving but do something in order to cope with the problem", whereas 48% indicated that they "would stop driving and do something in order to cope with the problem". Fortunately, only the 0.53% of the respondents answered that they would "Ignore the problem and continue driving". These participants were excluded from any further analysis.

Table 1. Percentage frequencies of night-time driving in the last 6 months for male and female young drivers.

% within	NEVER	Up to2 times a month	2-4 times a month	More than 4 times a month
Male	8.6%	14.8%	12.7%	63.9%
Female	21.5%	22.7%	18.2%	37.6%

Table 2. Linear regression results

a	B	t	Sig.	R^2	
Sex	0.07	2.33	0.02	0.04	$F_{(3, 1105)}$=9.7; $p<0.001$
Night-time driving frequencies	-0.12	-3.88	0.00		
Sleepiness at wheel	0.14	4.56	0.00		

[a] Dependent variable: Perceived risk of a car accident due to sleepiness.

The behaviors that they would adopt in order to cope with the problem of excessive sleepiness when driving at night are reported in Figure 1. The main behaviors they would adopt after deciding to continue driving were: "Open the windows" (18.10%), "Increase the effort and attention to driving" (13.25%), and "Increase radio volume "(12.75%). On the contrary, after deciding to stop driving, the main behaviors drivers would adopt were: "Take a coffee" (13.20%), "Wash one's own face" (10.19%), and "Take a nap"(7.7%).

The logistic regression analysis (Table 3) showed that the risky decision to continue driving during a sleepiness episode was positively and linearly related to the frequency of night-time driving ($Wald_{(3)}$= 22.34, p<.001). In particular, as illustrated in Table **3**, those who had driven at night "more than 4 time a month" were significantly more likely to choose to continue driving instead of stopping [ExpB = 2.41; p < .001] compared to the other participants. Furthermore, and not surprisingly, there also was a linear relation between the attitude to continue driving while perceiving sleepiness and the frequency of sleep attacks at the wheels ($Wald_{(2)}$= 7.71, p=.02).

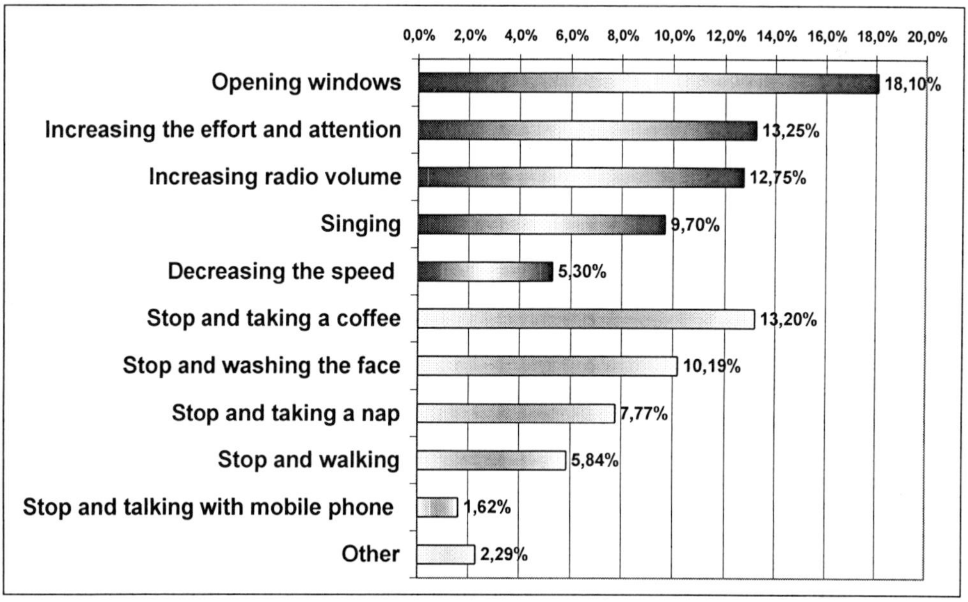

Figure 1. Frequencies of reported coping strategies for excessive sleepiness while driving.

Table 3 Logistic regression results

a	B	Wald	df	Sig.	Exp(B)
Sex	0.02	0.02	1	0.892	1.018
Night-time driving		22.34	3	0.000	
1-2 times	0.58	7.74	1	0.005	1.791
3-4 times	0.61	7.60	1	0.006	1.833
>4 times	0.88	22.15	1	0.000	2.411
Sleepiness at wheel		7.71	2	0.021	
1-3 times	0.16	1.23	1	0.267	1.171
> 3 times	0.64	7.41	1	0.006	1.892

[a] Dependent variable: Continue to drive while impaired by sleepiness

Conclusion

Our results indicated that the frequency of night-time driving influences risk perception and worries: the perceived likelihood of having a sleep-related car crash is lower for those who report to drive at night more frequently. Several studies indicate a correlation between the risk perception and the frequency of involvement in risky behaviors among adolescents: adolescents who engage in risky activities frequently attribute a low level of risk to those activities (Hampson et al., 2001; Gullone & Moore, 2000). This negative association is often interpreted within a cognitive approach in which unrealistic optimism (e.g. Weinstein, 1980) is enhanced by the frequency of adopting risky behaviors without experiencing their negative effects. In relation to novice drivers, Groeger (2000) pointed out that the illusion of control increases with driving experience. Driving under risky conditions without experiencing negative consequences can stimulate unrealistic optimism and an excess of self-confidence, which is typical of young drivers (Gregersen & Bjurulf, 1996; Rothengatter, 2002). On the other hand, our data indicated that the level of concern about sleep-related car crashes increases in young people who had experienced sleepiness-impaired driving. This experience could be conceived as a realistic feedback indicating that a situation in which demands exceed capability has been encountered, lowering the over-confidence bias. Unfortunately, some data show that individuals are not always subjectively aware of the decrease of performance due to sleep pressure (Devoto et al., 1999). Past experience of night-time driving also affects decisions concerning the behavior that could be adopted to cope with drowsiness at the wheels. The willingness to stop the car in order to cope with the problem of excessive sleepiness is less common among drivers who drive more frequently at night. In an attempt to continue to drive, various strategies are implemented, but many of them have been found to be ineffective. Reyner and Horne (1998) showed that the real effectiveness of such strategies, such as opening the window to get fresh air or increasing the volume of the car radio, was very low. Moreover, other authors (Dinges & Kribb, 1991; Dinges, 1992) showed that when there is a great need for sleeping, it is not possible to counteract the negative effects of sleep deprivation on performance by will power alone. With regards to the intake of caffeinated drinks, its effectiveness in alerting one's vigilance and enhancing performance varies according to one's individual level of caffeine tolerance (Balkin et al., 2004) and to one's individual resistance to sleep loss and circadian de-synchrony (Van Dongen, 2004). With regards to the effectiveness of strategies such as singing or talking on one's mobile phone, it is likely that the gains made to counteract sleepiness are vastly outweighed by the loss in terms of lack of concentration. People who refer they would certainly stop show greater awareness of the problem, but certain strategies, such as washing one's face or stretching one's legs, may reasonably be regarded as being scarcely effective.

In our opinion, the results of the present study could have relevant practical implications if used within education-based strategies. The observed effects of experience of night-time driving and sleep impaired driving on the perception of risk suggest that young drivers should be allowed to gain realistic feedback concerning the dangers involved in driving at night without risking the dramatic consequences of a car accident. For example, Lucidi et al. (2002) carried out a demonstration aimed at allowing young drivers to experience realistic feedback from the nocturnal decrease of their performance in a vigilance task, performed at their school during day-time and night-time. Moreover, the lack of knowledge about the real efficacy of the different coping strategies used to counteract sleepiness at the wheels could be overcome by specific informative and educational interventions in young drivers. However, just knowing which are the effective strategies to counteract sleepiness at the wheels and when to use them is not enough in the absence of proper environmental conditions. For these reasons, educational and informative strategies should be matched to structural intervention such as, for example, increasing the availability of rest areas and coffee points along roads and highways.

ACKNOWLEDGMENT

Research supported in part by a MUR PRIN 2006 grant to Cristiano Violani.

REFERENCES

Akerstedt, T., Czeisler, C. A., Dinges, D. & Horne, J. A. (1994). Accidents and sleepiness: a consensus statements. *Journal of Sleep Research, 4,* 195.

Akerstedt, T. & Kecklund, G. (2001). Age, gender and early morning highway accidents. *Journal of Sleep Research, 10,* 105–110.

Balkin, T. J., Kamimori, G. H., Redmond, D. P., Vigneulle, R. M., Thorne, D. R., Belenky, G. & Wesensten, N. J. (2004). On the importance of countermeasures in sleep and performance models. *Aviation Space and Environmental Medicine, 75 (Suppl. 3),* A155–A157.

Carskadon, M. A. (2002). Risks of driving while sleepy in adolescents and young adults. In: M.A. Carskadon, (Ed.), *Adolescent Sleep Patterns: Biological, Social, and Psychological Influences* (pp. 148–158). New York, NY, US: Cambridge University Press.

Connor, J., Norton, R., Ameratunga, S., Robinson, E., Civil, I., Dunn, R., Bailey, J. & Jackson, R. (2002). Driver sleepiness and risk of serious injury to car occupants: population based case control study. *British Medical Journal, 324 (7346),* 1125.

Deery, H. A. (1999). Hazard and risk perception among young novice drivers. *Journal of Safety Research, 30 (4),* 225–236.

DeJoy, D. M. (1992). An examination of gender differences in traffic accidentrisk perception. *Accident Analysis and Prevention, 24 (3),* 237–246.

Devoto, A., Lucidi, F., Violani, C. & Bertini, M. (1999). Effects of different sleep reductions on daytime sleepiness. *Sleep, 22 (3),* 336–343.

Dinges, D. (1992). Probing the limits of functional capability: the effects of sleep loss on short duration task. In: R. Broughton & R. Ogilvie (Eds.), *Sleep, Arousal and Performance* (pp. 176–188). Boston: Birkhauser–Boston Inc.

Dinges, D. (1995). An overview of sleepiness and accidents. *Journal of Sleep Research, 4 (2),* 4–14.

Dinges, D. & Kribbs, N. B. (1991). Performing while sleepy: effects of experimentally induced sleepiness. In: T. H. Monk (Ed.), *Sleep, Sleepiness and Performance* (pp. 97–128). Chichester, UK: John Wiley and Sons Ltd.

Folkard, S. (1997). Black times: temporal determinants of transport safety. *Accident Analysis and Prevention, 29 (4),* 417–430.

Fuller, R. (2000). The task–capability interface model of the driving process. *Recherche Transports Sécurité, 66,* 47–59.

Gregersen, N. P. (1996). Young drivers' overestimation of their own skill—An experiment on the relation between training strategy and skill. *Accident Analysis and Prevention, 28,* 243–250.

Gregersen, N. P. & Bjurulf, P. (1996). Young novice drivers: towards a model of their accident involvement. *Accident Analysis and Prevention, 28 (2),* 229–241.

Groeger, J. A. (2000). *Understanding Driving.* Hove, England: Psychology Press.

Gullone, E. & Moore, S. (2000). Adolescent risk-taking and the five-factor model of personality. *Journal of Adolescence, 23,* 393–407.

Hampson, S. E., Severson, H. H., Burns, W. J., Slovic, P. & Fisher, K. J. (2001). Risk perception, personality factors and alcohol use among adolescents. *Personality Individual Differences, 30,* 167–181.

Harre′, N. (2000). Risk evaluation, driving, and adolescents: A typology. *Developmental Review, 20,* 206–226.

Horne, J. A. & Reyner, L.A. (1995a). Sleep related vehicle accidents. *British Medical Journal, 310,* 565–567.

Horne, J. A. & Reyner, L. A. (1995b). Driver sleepiness. *Journal of Sleep Research, 4 (2),* 23–29.

Jennings, J. R., Monk, T. H. & Van der Molen, M. W. (2003). Sleep deprivation influences some but not all processes of supervisory attention. *Psychological Science, 14 (5),* 473–479.

Johnson, L. C. (1982). Sleep deprivation and performance. In: W.B. Webb (Ed.), *Biological Rhythm, Sleep, and Performance* (pp. 111–138). Chichester: Wiley.

Kuiken, M. J. & Twisk, D. A. M. (2001*). Safe driving and the training of calibration: A literature review* (Report R2001-29). Leidschendam: Institute of Road Safety Research.

Langolis, P. H., Smolensky, M. H., Hsi, B. P. & Weir, F. W. (1986). Temporal patterns of reported single-vehicle car and truck accidents in Texas, USA, during 1980-1983. *Chronobiology International, 2,* 131–146.

Leger, D. (1995). The cost of sleepiness: a response to comments. *Sleep 18,* 281–284.

Lucidi, F., Devoto, A., Bertini, M., Braibanti, P. & Violani, C. (2002). The effects of sleep debt on vigilance in young drivers: an education/research project in high schools. *Journal of Adolescence, 25,* 405–414.

Lyznicki, J. M., Doege, T. C., Davis, R. M. & Williams, W. A. (1998). Sleepiness, driving and motor vehicle crashes. *Journal of the American Medical Association 279,* 1908–1913.

McConnel, C. F., Bretz, K. M. & Dwyer, W. O. (2003). Falling asleep at the wheel: a close look at 1269 fatal and serious injury-producing crashes. *Behavioral Sleep Medicine, 1(3),* 171–183.

Mitler, M. M., Carskadon, M. A., Czeisler, C. A., Dement, W. C., Dinges, D. F. & Graeber, R. C. (1988). Catastrophes, sleep and public policy: consensus report. *Sleep, 11,* 100–109.

Pack, A. I., Pack, A. M., Rodgman, E., Cucchiara, A., Dinges, D. F. & Schwab, C. W. (1995). Characteristics of crashes attributed to the driver having fallen asleep. *Accident Analysis and Prevention, 27 (6),* 769–775.

Philip, P., Ghorayeb, I., Stoohs, R., Menny, J. C., Dabadie, P., Bioulac, B. & Guilleminault, C. (1996). Determinants of sleepiness in automobile drivers. *Journal of Psychosomatic Research, 41 (3),* 279–28.

Prokop, O. & Prokop, L. (1955). Ermuhdung und einschlafen am steurer. *Zbl. Verkehrsmed, 1,* 19–30.

Reyner, L. A. & Horne, J. A. (1998). Evaluation of «in car» countermeasures to driver sleepiness: cold air and radio. *Sleep, 21,* 46–50.

Rothengatter, T. (2002). Drivers' illusions—no more risk. *Transportation Research Part F: Traffic Psychology and Behaviour, 5,* 249–258.

Sivak, M., Soler, J., Traenkle, U. & Spagnhol, J. M. (1989). Cross-cultural differences in driver risk-perception. *Accident Analysis and Prevention, 21 (4),* 355–362.

Summala, H. (1997). Hierarchical model of behavioural adaptation and traffic accidents. In: J. A. Rothengatter & E. Carbonell Vaya (Eds.), *Traffic and Transport Psychology: Theory and Application.* Oxford: Pergamon.

Summala, H. & Mikkola, T. (1994). Fatal accidents among car and truck drivers: effects of fatigue, age and alcohol consumption. *Human Factor, 36,* 315-326.

Van Dongen, H. P. A. & Balkin, T. J. (2004). Commentary on modeling performance and alertness: the QinetiQ approach. *Aviation Space and Environmental Medicine, 75 (Suppl. 3),* A104–A105.

Webb, W. B. (1995). The cost of sleep-related accidents: a reanalysis. *Sleep, 18 (4),* 276–280.

Weinstein, N. D. (1980). Unrealistic optimism about future life events. *Journal of Personality and Social Psychology, 39 (5),* 806–820.

Zomer, J. & Lavie, P. (1990). Sleep-related accidents-when and who? In: J. A. Horne (Ed.), *Sleep '90* (pp. 448–451). Bochum: Pontenagel Press.

Zulley, J., Cronlein, T., Hell, W. & Langwiender, K. (1994). Fatal highway accident mainly caused by fallen asleep. In: T. Akerstedt, & G. Kecklund (Eds.), *Work hours, Sleepiness and Accidents.* Stockholm: IPM and Karolinska Institute.

In: Accidents: Causes, Analysis and Prevention
Editors: Henri Bédard and Geraud Delashmit

ISBN: 978-1-60741-712-5

Chapter 7

FEATURE ANALYSIS OF HIGHWAY ACCIDENT DURATION IN TAIWAN

***Ying Lee*[*1] and *Chien-Hung Wei*[†2]**

[1]Department of Hospitality Management, Ming Dao University
No.369, Wen-Hua Road, Pee-Tow Township, Chang-Hua County, Taiwan
[2]Department of Transportation and Communication Management Science, National Cheng Kung University, No.1, Ta-Hsueh Road, Tainan city, Taiwan

ABSTRACT

Highway congestion is seriously affected by traffic accidents. The stochastic nature associated with an accident causes significant difficulties in accident duration forecast, travel time prediction and route guidance. Therefore it is vital to gain an insight into the characteristics of accidents. In this study we analyze the accident database from the Police Radio Station in order to find useful clues with any possible data feature. Focus is mainly on the impact of accident duration on traffic congestion. Further studies could refer this study results to process the model inputs for model development.

Keywords: Data characteristic, Accident duration, Traffic data, Highway.

1. INTRODUCTION

Accidents are major contributors to traffic congestion. Extra travel time and fuel consumption induced by congestion, both recurring and occasional, has been estimated as totaling $35 billion for 50 large U.S. urban areas in 1988 (Small, 1997). Freeways have borne the brunt of an increasing number of vehicles, commensurate with congestion and accident rates. If relevant traffic data can be processed and analyzed effectively, forecast accident duration information could be used to reduce the uncertainty faced by the traveling public.

---

* Corresponding author: Fax: +886-4-8879035; E-mail: yinglee1017@gmail.com
† Fax: +886-6-2753882; E-mail: louiswei@mail.ncku.edu.tw

This study focuses on analyzing the relationship between the data feature and the accident duration. Section 2 brief reviews the accident duration research on the model inputs. Section 3 presents the accident data source and the definition of accident duration. Section 4 discusses the relationship between the data feature and the accident duration. The final section presents concluding remarks and suggestions for future research.

2. Literature Review of Accident Duration

Garib et al. (1997) employed the multiple regression analysis to predict the clearance time of the accident. The model shows that 81% of all variations in accident duration can be predicted by the number of lanes affected, number of vehicles involved, truck involvement, time of day, police response time, and weather conditions.

Nam and Mannering (2000) applied a hazard-based analysis by deriving three accident duration models from occurrence, notification and the arrival of emergency services. Their results show that a wide variety of factors significantly affected accident duration, including: the month, rain, geographic information and the extent of injuries (if any). Regarding notification, the significant variables are temporal characteristics (peak time; night; Saturday; Sunday) and geographic information. It is generally the case that the scene of an accident is attended in the first instance by a different agency on separate occasions. The third model, forecasting how much time will elapse till the rescue agency arrives, considers the lead agency as the model input.

Wei and Lee (2005) developed the accident duration model by using techniques of data fusion. This model forecasts duration from the point of accident notification. In accordance with factor sensitivity analysis, the significant factors can be identified. It follows that time of day, number of vehicles and vehicle type involved in accident and road geometry, are significant factors for the duration model.

In 2007, Wei and Lee extended the accident duration forecasting study by incorporating the point of notification through to road clearance of the accident scene. Model input variables include incident characteristics, traffic data from VD, time relationship, space relationship, and geometry characteristics.

3. Accident Data Source

In Taiwan, The most reliable source of incident data is provided by the Police Radio Station and updated every 180sec. In this incident database, the incident includes car accident, traffic barrier, and recurrent congestion. Each incident is shown repeatedly in the database until it is cleared away from the traffic lane occupied. Table 1 shows the incident data in original form. The time, the incident type and incident location are recorded. The duration time of each incident can be calculated from its first and last record. This study only chooses the car accident data for analysis. Through data filtering process, this study sieves out the accident data from incident database for discussion.

Table 1. Examples of incident data in original form

Date	2005/1/31	2005/1/31	2005/1/31
Time	23:23:59	23:28:53	23:29:48
Location	Taipei city	No. 1 national freeway	No. 1 national freeway
Direction		Southern bound	Northern bound
Beginning of accident site (KM)	N/A	51.8	52.9
Ending of accident site (KM)	N/A	N/A	N/A
Incident type	Accident	Maintenance	Maintenance
Incident situation	Collision with two cars at cheng-de bridge	Road maintenance at shoulder and outside lane	Road maintenance at shoulder and outside lane
Provider	Traveler	Maintenance staff	Maintenance staff

N/A: Not available

Table 2. The Duration in Five Conditions of Occupied Lanes

Occupied lanes	Shoulder	Outer	Inner	Outer & Shoulder	Inner & Outer
Mean of duration (Sec)	1840	2865	2990	5333	6069
Std. of duration (Sec)	123	362	331	886	1395

This study selected the southern part of the No.1 national freeway in Taiwan, which is from An-ding interchange to Lu-jhu interchange, as the experiment site. This double-lane site has a length of 27.2 km, 1 toll station, 3 interchanges and 2 system interchange, as shown in Figure 1. On the southbound segment, there are seven VD, and speed limit is 100 km/hr. The raw data are collected for 24 hours a day over a 6-month period from December 2004 to September 2005 (excluding February, March, April, May 2005). The time period which covers weekdays/weekends and peak/non-peak hours adequately reflects real traffic conditions.

4. Feature Analysis of Accident Duration

Accident duration could be affected by the following factors, lane occupied by accident, number of vehicles and vehicle type involved in accident, temporal relationship, spatial relationship, geometry characteristics and traffic condition.

4.1. Occupied Lane

When the accident is reported, lane occupied by accident is one of the important features which are easy to be described. Table 2 and Figure 2 show five conditions of the occupied lanes. When only the inner lane or outer lane is occupied in an accident, the mean of duration is roughly the same. We merge these two conditions into one and consider it one lane occupied. Based on the same reason, the conditions of inner & outer lanes occupied and outer lane & shoulder occupied are merged and declared two lane occupied.

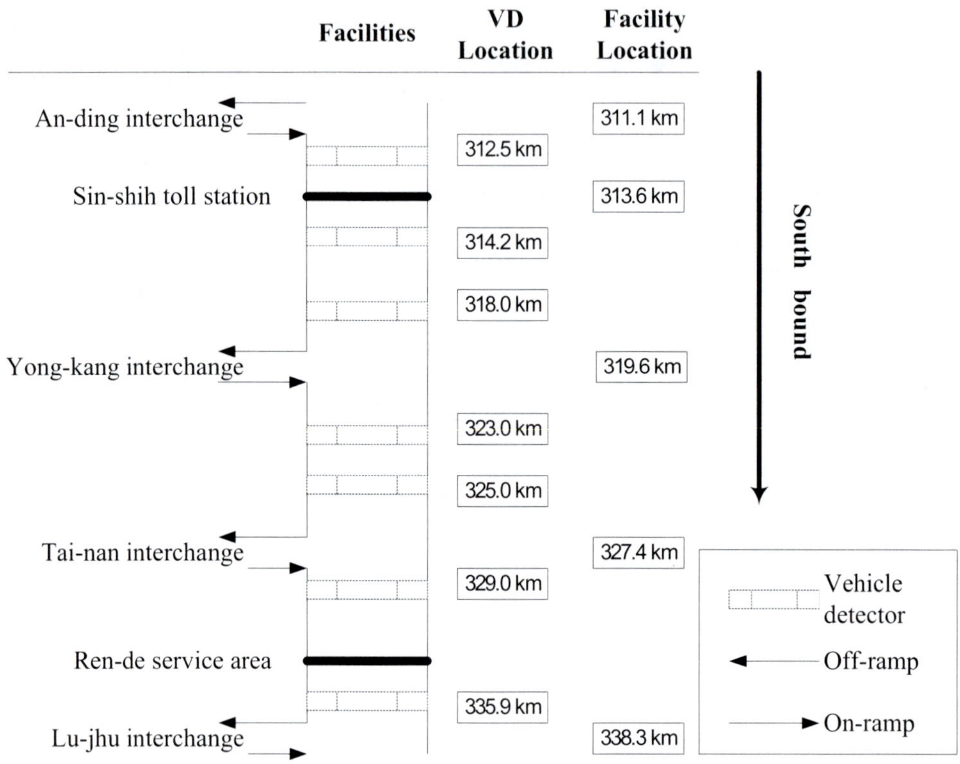

Figure 1. Freeway layout from An-ding to Lu-jhu (South bound)

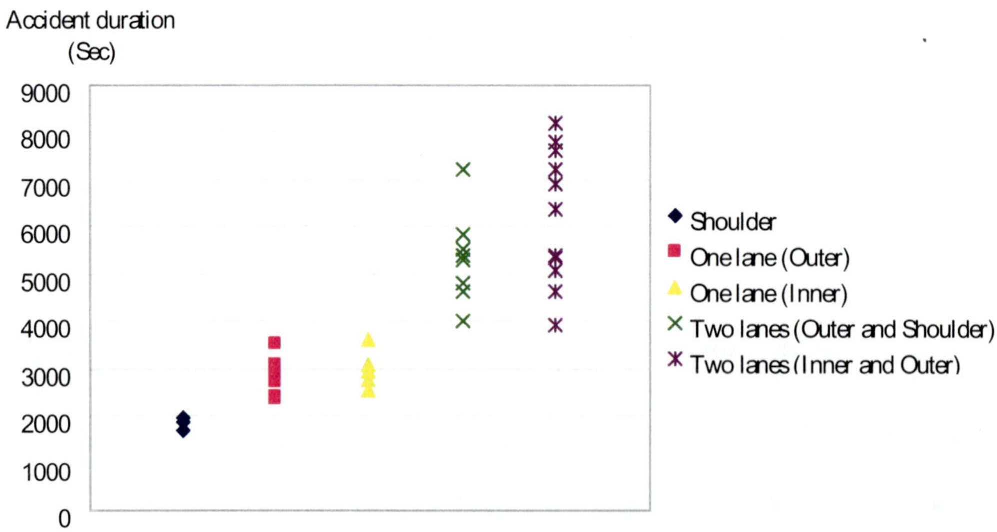

Figure 2. The relationship between accident duration and occupied lane

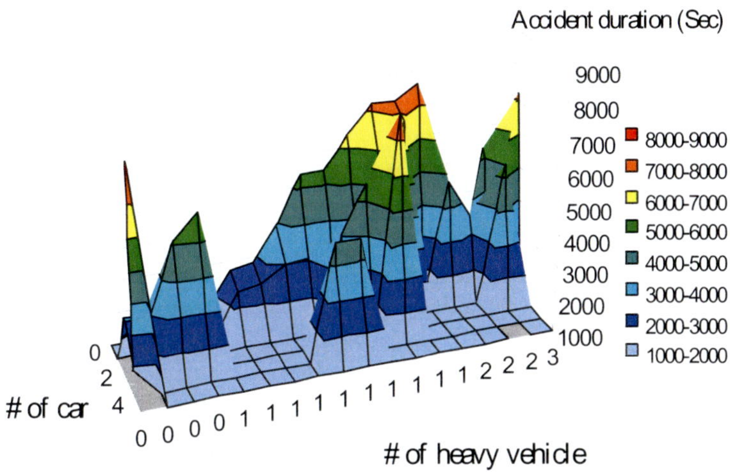

Figure 3. The relationship between accident duration and number of vehicles

4.2. Number of Vehicles and Vehicle Type Involved

The relation among the number of vehicles, vehicle type and the duration is displayed in Figure 3. In the Figure, vertical coordinate the number of passenger cars in an accident, horizontal coordinate the number of heavy vehicles (truck, bus and trailer) in an accident and the color represents the accident duration. Each accident is plotted in the Figure according to the number of passenger cars and heavy vehicles involved. As the Figure 3 shows the accidents result in longer accident duration mostly involve heavy vehicles.

4.3. Temporal Relationship

Figure 4 shows the relationship between the accident duration and time of day. The time of accident notification is related to the difference of accident duration between the daytime from 7:00 a.m. to 6:30 p.m. and the nighttime from 6:30 p.m. to 7:00 a.m. The VD data are accumulated and recorded every 300sec for each lane. There probably exists a time lag between the time of an accident notification and the recording time of VD data. The time lag may be considered as an input to represent the possible variation of traffic situation between the time of accident notification and the recording time of VD data.

4.4. Spatial Relationship

The distance between an accident location and VD site presents likely variation of traffic situation. Generally speaking, the greater the distance, the longer the time is required to detect the traffic flow data influenced by an accident. When an accident occurs, rescuers arrive at the accident location from neighboring interchanges. The distance between the accident and the neighboring interchanges is used to represent the time that rescuers take to arrive at an accident location.

Wei and Lee [2005] had disclosed the relationship between the accident duration and the distance from the accident location to the interchange.

The distance from the accident location to the near by interchange is displayed in Figure 5. Vertical coordinates the distance between the accident location and the nearest upstream interchange, horizontal coordinates the distance between the accident location and the nearest downstream interchange. The accident duration are shown by the different color. In Figure 5, every point represents an accident location and corresponding accident duration. When the distance from the accident location to the interchange is between 4 and 7 km, the accident duration usually is longer. These accidents are marked with the circles in Figure 5.

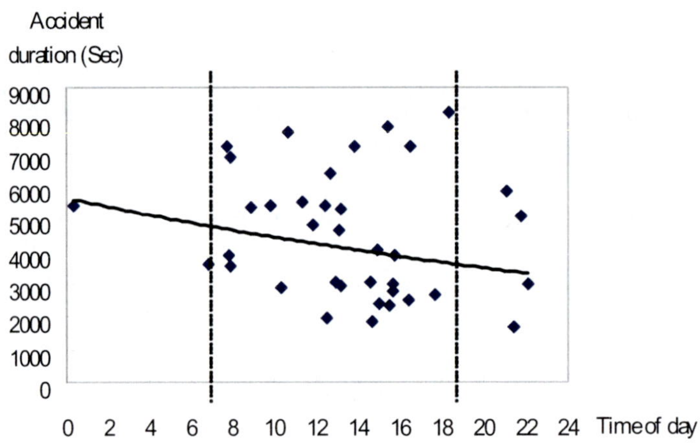

Figure 4. The relationship between accident duration and time of day

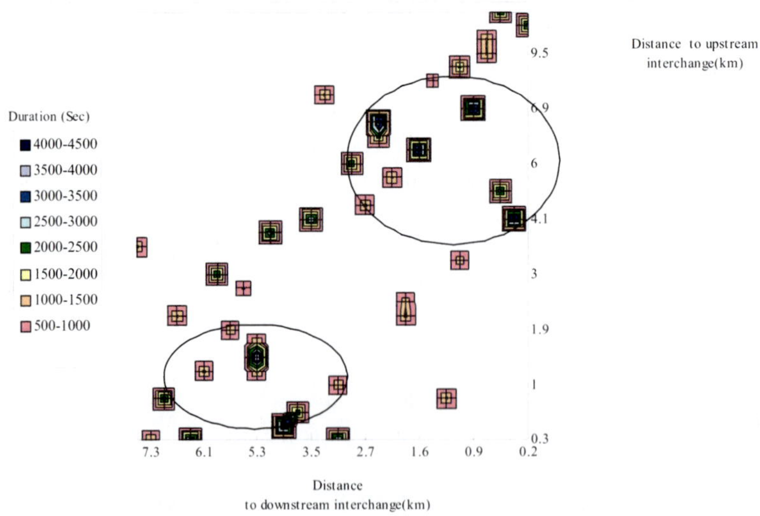

Figure 5. The Relationship between the Distance from the Interchange and the Duration

4.5. Geometry Characteristics

Highways have several kinds of geometric characteristics, and similar accidents may result in different durations in different geometries. Accident locations can be classified as interchange area, service facility area, and other types of areas. If an interchange is close to an accident and VD sites, VD data reflecting the traffic situation after the occurrence of an accident will be affected by that interchange. This is because the existence of the interchange between the accident and the VD in neighboring can explain the influence of the interchange on the VD data. A similar reason applies for considering the existence of a service facility between the accident and the neighboring VD [Wei and Lee, 2005].

If the accident locates within 1.5 in the front and rear of the interchange or the service facility km, the accident belongs to the corresponding area. The service facility area includes the toll station and the driver service area. The relationship between three classified areas and the duration is shown in Table 3 and Figure 6. Among these three areas, the distribution of the duration is different from each other.

4.6. Traffic Data

With respect to an accident location, traffic data collected by the first and the second VD at upstream and the first VD at downstream are used to present the traffic situation in the front and back of an accident. According to the average spacing between the neighboring VDs in Taiwan, considering two upstream VDs can mostly cover the roadway segments affected by an accident..

Table 3. The Duration in three kinds of highway area

Highway	Interchange area	Service facility area	Others
Mean of duration (Sec)	2403	1980	1790
Std. of duration (Sec)	993	382	655

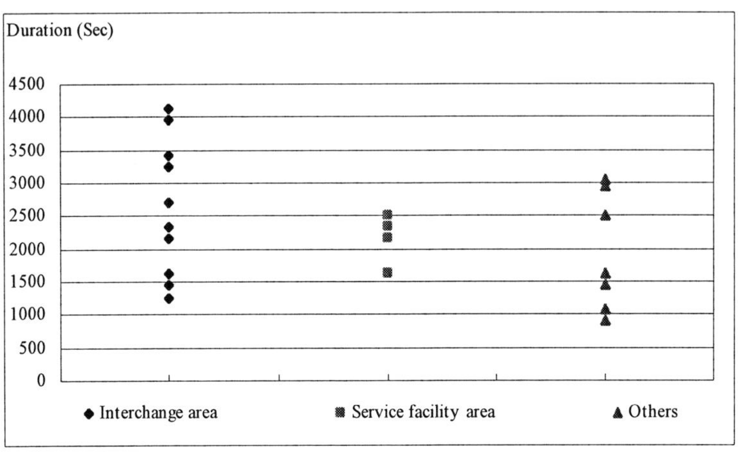

Figure 6. The Relationship between the Highway Geometry and the Duration

5. CONCLUSION

The uncertainty of travel time increases when the travel conditions are unknown, although with advanced traffic data collection techniques, traffic data can be collected much more easily. Traffic patterns could be adequately characterized if the data could be processed and analyzed effectively, thereby providing useful information such as travel time information for those planning to travel (D' angelo et al., 1999; Chen & Chien, 2001; Krikke, 2002). Many studies have discussed travel time forecast issues, although few of them have considered the influence of incidents in their models (Wei & Lee, 2003; Zhang & Rice, 2003).

This study analyzes the accident database operated by the Police Radio Station in order to find useful clues with any possible data feature. Focus is mainly on the duration of accident impacts on the traffic. Further studies, such as accident duration forecast and travel time prediction, could refer to these findings to process the model inputs. The forecast accident duration and predicted travel time information could be used to reduce the uncertainty faced by the traveling public.

ACKNOWLEDGMENTS

This study is based on research projects sponsored by the National Science Council of Taiwan, R.O.C. [NSC93-2218-E-006-094][NSC97-2221-E-451-010-MY2]. The authors gratefully acknowledge the data providers for this research, including Taiwan Area National Freeway Bureau and Police Radio Station. The helpful comments of Dr. Steven Chien at the New Jersey Institute of Technology, USA, are sincerely appreciated.

REFERENCES

Chen, M. & Chien, I. J. (2001). Dynamic freeway travel time prediction using probe vehicle data: link-based vs. path-base. *Transportation Research Record 1768*, TRB, National Research Council, Washington, D. C., 157-161.

D'angelo, M. P., Al-Deek, H. M. & Wang, M. C. (1999). Travel time prediction for freeway corridors. *Transportation Research Record 1676*, TRB, National Research Council, Washington, D. C., 184-191.

Garib, A., Radwan, A. E. & Al-Deek, H. (1997). Estimating magnitude and duration of incident delays. *Journal of Transportation Engineering., 123(6),* 459-466.

Krikke, R. (2002). Short-range travel time prediction using an artificial neural network. *Proceedings of the ninth ITS World Congress (CD-ROM)*, Chicago, USA.

Nam, D. & Mannering, F. (2000). An exploratory hazard-based analysis of highway incident duration. *Transportation Research Part A., 34(2),* 85-102.

Wei, C. H. & Lee, Y. (2005). Appling data fusion techniques to traveler information services in highway network. *Journal of the Eastern Asia Society for Transportation Studies. 5,* 2457-2475.

Wei, C. H. & Lee, Y. (2007). Data fusion and feature composition approach to sequential accident duration forecasting. *Journal of the Eastern Asia Society for Transportation Studies 7,* 2546-2560.

Wei, C. H. & Lee, Y. (2007). Sequential forecast of incident duration using artificial neural network models. *Accident Analysis and Prevention 39(5)*, 944-954.

Zhang, X. & Rice, J. A. (2003). Short-term travel time prediction. *Transportation Research, Part C, 11(3-4),* 187-210

In: Accidents: Causes, Analysis and Prevention
Editors: Henri Bédard and Geraud Delashmit ISBN: 978-1-60741-712-5

Chapter 8

Information Processing, Personality, and Demographic Variables as Predictors of Crashes and Moving Violations

Winfred Arthur[1] and Eric Anthony Day[2]
[1]Department of Psychology, Texas A&M University, 4235 TAMU, College Station, TX 77843-4235, USA
[2]Department of Psychology, The University of Oklahoma, 455 W. Lindsey, Room 705, Norman, OK 73019-2007, USA

Abstract

The objective of the present research was to undertake a comparative and combined assessment of the three commonly used predictor categories of driving outcomes, namely demographic, information processing, and personality variables. Research participants (Study 1: n = 153; Study 2: n = 335) completed a predictor battery and a self-report measure of motor vehicle crashes and moving violations. Using a series of hierarchical regression analyses, the obtained results generally suggest that, consistent with the conceptual need to recognize the role of the "total" individual in driving-related behaviors, the use of multiple predictors from multiple predictor categories in the prediction of driving outcomes is empirically meaningful. As would be expected, this approach accounted for substantially more variance in the criteria than either any single predictor or individual predictor category would by itself. In addition, the findings particularly highlighted the importance of including multiple personality variables in models of driving outcomes. The implications of these findings and suggestions for future research are discussed. Information Processing, Personality, and Demographic Variables as Predictors of Crashes and Moving Violations.

Introduction

The objective of the present research was to demonstrate that because driving is a complex task whose performance is influenced by multiple individual difference predictor categories, a theoretical and empirical combination of these predictor categories results in incremental prediction and validity over and above any single predictor. This view is based on the distinction between errors and violations as anomalous behaviors (Arthur, Barrett, and Alexander, 1991; Parker, Reason, Manstead, and Stradling, 1995; Reason, 1990; Reason, Manstead, Stradling, Baxter, and Campbell, 1990) and the different roles they play in driving outcomes. Specifically, because errors result from failures in information processing, and violations in contrast, are more motivational, value, and attitudinal-oriented, different individual difference variables will capture different elements of the driving behavior criterion space. Thus, although the nature of our driving data did not allow us to map on specific crash-types to specific predictors, the conceptual basis of this reasoning and the complexity of driving behavior leads us to posit that a combination of predictor-types will more effectively capture the driving criterion space than any single predictor by itself. Consequently, a comparative and combined assessment of three commonly used predictor categories of driving outcomes, namely demographic, information processing, and personality variables (Arthur et al., 1991; Elander, West, and French, 1993; Hansen, 1989) was undertaken.

Demographic Variables

Commonly studied demographic variables in the driving literature can be conceptualized as the exposure factors that facilitate the operation of other variables that result in crashes and tickets instead of being causal variables themselves. Examples of these exposure variables include number of years driving, type of driving (day or night, work or personal, rural or urban), and amount of driving (e.g., miles driven per week; National Safety Council, 1997).

As demographic variables, age and sex have probably stimulated the most interest in the motor vehicle crash research literature. Although there are unequivocal age differences in crash involvement (Butler, 1982; Insurance Institute for Highway Safety, 1990; Massie, Campbell, and Williams, 1995; National Highway Traffic Safety Administration, 1989; National Safety Council, 1997), it appears that different factors contribute to crashes for different age groups (Elander et al., 1993).

Research has also indicated that different age groups are involved in different types of crashes. Evans (1991) has shown that driving fatalities among older drivers are typically the result of side-impact collisions, whereas driving fatalities among younger drivers are more likely to result from rollover crashes. It could also be argued that age, in general, is a proxy variable for exposure; that is, it covaries with the number of driving years and, therefore, opportunities for crashes. It is also conceivable that age could simply be serving as a proxy variable for the intrinsic attributes required for driving (Arthur, Fuentes, and Doverspike, 1990). In sum, although there appears to be a relationship between age and crash involvement, the relationship is not fully understood.

With respect to sex differences in driving outcomes, males tend to have higher involvement rates in fatal car crashes, whereas females have a higher involvement rate for all crashes (National Safety Council, 1997; Massie et al., 1995). Although age and sex differences in crash involvement rates are interesting, it has been argued that age and sex are proxies for other predictor variables that are more directly related to crash involvement (e.g., exposure, risk-taking behavior, visual acuity, and personality), and that research should focus on these more direct links (cf. Arthur et al., 1990; McEvoy and Cascio, 1989). Thus male/female and age differences in driving outcomes are best conceptualized as differences in information processing, and motivations, attitudes, and values—that is, personality.

Information Processing Variables

Some of the most commonly investigated information processing variables in the driving literature include visual and auditory attention, perceptual style, choice and complex reaction time, and to a lesser extent, cognitive ability (Arthur et al., 1991; Arthur and Doverspike, 2001). From an information processing perspective, complex perceptual information processing tasks, such as driving, call for perceiving, identifying, processing, and adequately responding to pertinent information (e.g., traffic lights, signs, pedestrians, other vehicles, etc.) in the environment. Errors (slips, lapses, or mistakes) in the processing of this information could result in crashes. Thus, on a conceptual and theoretical basis, one would expect individual differences on information processing variables to be related to crash involvement. This relationship has been extensively demonstrated. Arthur et al.'s meta-analysis obtained moderate to marginally favorable results for selective attention, perceptual style, and cognitive ability. Additional primary studies (e.g., Arthur and Doverspike, 1992; Arthur, Strong, and Williamson, 1994; Robinson and Bownas, 1993) have obtained similar results. These findings are consistent with Ball and Owsley's (1991) identification of "useful field of view" as a predictor of crash involvement, particularly in older adults. Ball, Sloane, Roenker, and Bruni (1993) have identified speed of visual information processing, inability to ignore distractors, and inability to divide attention as factors related to crashes. Nevertheless, the size of the effects for information processing variables as predictors of crashes are generally modest, and it is not uncommon to find recent primary studies that fail to find statistically significant relationships between information processing variables and crash involvement (e.g., Martín and Estévez, 2005).

Personality Variables

The personality/driving literature has a long history and a correspondingly large number of personality variables that have been investigated in terms of their ability to predict driving outcomes. Our review is limited to a few of the more prominent variables. The role of personality variables (in contrast to information processing or ability) reflects the choices that drivers make (Elander et al., 1993). These choices reflect individual differences in attitudes, values, beliefs, and motivations. Consequently, personality variables are conceptually related to crashes through the choices that individuals make in terms of intentional deviations from safe driving practices and behaviors (Parker, Manstead, Stradling, and Reason, 1992; Parker, Stradling, and Manstead, 1996). One personality variable that has been extensively researched in the crash involvement literature is extraversion (Eysenck and Eysenck, 1975). Several studies (e.g., Fine, 1963; Lajunen, 2001; Loo, 1979) have found a significant relationship between extraversion and crash involvement. However, a number of studies (e.g., Craske, 1968; Dahlen and White, 2006; Singh, 1978; Wilson and Greensmith, 1983) have failed to support these findings. Reflective of the mixed nature of these findings, Arthur and Graziano (1996) obtained a significant correlation between extraversion and a self-report measure of number of crashes in one sample, but failed to replicate it in another with total crashes or number of tickets (i.e., moving violations).

Historically, the evidence for systematic links between personality variables and crash involvement was relatively poor until the advent of robust models of personality, like the five-factor model (Digman, 1990; McCrae and Costa, 1987; McCrae and John, 1992). Before the five-factor model, personality variables seemed to be characterized by a lack of standardization unlike variables in other predictor categories (e.g., information processing; Arthur et al., 1991). It was not uncommon for personality researchers to use unstandardized, ad hoc questionnaires, inventories, and tests developed specifically for their particular studies (e.g., Clark, 1976; Schuster and Guilford, 1964). As an attempt to address this concern, Arthur and Graziano (1996) used the five-factor model of personality to investigate the relationship between conscientiousness and crash involvement. Arthur and Graziano (1996) demonstrated that individuals who rate themselves as more self-disciplined, responsible, reliable, and dependable are less likely to be involved in crashes than those who rate themselves lower on these attributes. More recently, Bogg and Roberts (2004), in their meta-analytic review of conscientiousness and health-related behaviors, showed that the self-control and traditionalism facets of conscientiousness more strongly predicted risky driving compared to the industriousness, order, and responsibility facets. In general, extraversion and conscientiousness appear to be the five-factor traits most likely to be associated with driving outcomes. Evidence linking agreeableness, openness, and emotional stability to driving outcomes, particularly crashes, appears to be more inconsistent (cf. Dahlen and White, 2006; Lajunen, 2001).

The limited review presented above demonstrates that each of the predictor categories identified by Arthur et al. (1991) can be theoretically and conceptually related to driving outcomes, and furthermore, these relationships have been empirically demonstrated, albeit some more consistently than others. However, as noted by Arthur et al., these predictor categories have been conceptualized as though they were mutually exclusive. Thus, previous studies have typically looked at either demographic, information processing, or personality predictors with little or no attempt made to investigate how combinations of predictors across categories may improve the ability to predict driving outcomes. In fact, the use of multiple predictors is a relatively common practice in personnel selection where multiple predictors with low inter-correlations are used to enhance the ability to predict the job performance criterion of interest.

Consequently, viewing driving as a complex task whose performance is influenced by information processing and ability, and personality factors, as well as demographic and exposure variables, it seems reasonable to hypothesize that a theoretical and empirical combination of these perspectives may result in incremental prediction and validity over and above any single predictor. As noted by Hansen (1989), many constructs are relevant to driving behavior, and no single ability or skill will predict all aspects of driving. Furthermore, if crashes can result from *both* errors and violations, each of which has a different psychological basis (i.e., information processing versus personality, respectively), then one would expect the best prediction to be obtained from an approach that incorporates *both* categories of predictor variables. Consequently, the prevailing practice of using single predictors in the extant literature might be one of the primary reasons for the typically low and inconsistent validities obtained for the prediction of crash involvement (Guastello, 1993). Because the totality of the individual is involved in driving and subsequent crash- and ticket-related behaviors, it seems more appropriate to try to predict these behaviors using the gestalt of the individual instead of separate facets of their individuality.

In summary, using data from two studies, the objective of the present research was to undertake a comparative and combined assessment of the three commonly used predictor categories of driving outcomes. Demographic and exposure variables were used as control variables and the effects of information processing and personality variables, over and beyond these variables, were assessed. The criterion data used in both Study 1 and Study 2 were self-report measures of the total number of crashes that participants had been involved in and the total number of tickets (moving violations) they had. Due to the methodological problems associated with collecting self-report data of this sort, it was not possible to collect these data at the disaggregated level of violations and errors (i.e., slips, lapses, and mistakes).

STUDY 1: METHOD

Participants

Participants were 153 paid volunteers recruited from Texas A&M University and its community by an advertisement in the school newspaper and posted notices around campus. Because the data were collected as part of a larger all-male study on complex skill acquisition (in which participants were paid $75 for their participation), all participants in Study 1 were male. The mean age for the sample was 20.41 years (*SD* = 2.59). Participants' data were excluded if they reported visual impairments or if they did not have a valid driver's license. The mean number of years the study participants had been driving was 4.97 years (*SD* = 2.72).

Measures

A personality measure, based on the five-factor model, was used to operationalize personality. Four information processing variables were used in the present study. These were visual attention, cognitive ability, spatial processing speed, and spatial working memory. Two demographic variables—age and total number of years driving—were operationalized via self-reports on a driving behavior questionnaire. The measures used to operationalize each of the study variables are described below.

Goldberg's 100 Unipolar Markers (Goldberg, 1992). This is a standard measure of the dimensions of the five-factor model of personality. Using a 9-point scale (1 = extremely inaccurate, to 9 = extremely accurate), participants rate a list of 100 common human traits in terms of how accurately they describe themselves. Each factor is measured by 20 items and, for purposes of the present research, a participant's score on each of the five factors was the sum of their ratings on the 20 items that represent that factor after the appropriate items had been reverse scored. For self-assessments, internal consistencies of .90 (extraversion), .88 (agreeableness), .90 (conscientiousness), .84 (emotional stability), and .85 (openness) have been reported for the five factors (Goldberg, 1992). For the present study, internal consistencies ranged from .85 to .91.

Computer-Administered Visual Attention Test (CA-VAT; Arthur et al., 1994). The CA-VAT is a computer-administered and scored test of visual attention. The CA-VAT was constructed as an approximate visual counterpart to the Auditory Selective Attention Test (ASAT; Gopher and Kahneman, 1971; Mihal and Barrett, 1976). The CA-VAT in Study 1 was scored in terms of errors. Arthur et al. (1994) reported moderate convergent validity for the CA-VAT and ASAT (r = .25) and internal consistencies of .93 and .98. A test-retest reliability of .83 has also been reported for the CA-VAT (Strong, 1992). In addition, Arthur et al. (1994) provide predictive validity evidence for the CA-VAT. An internal consistency of .84 was obtained for the present study.

Figure Matrices Test (cognitive ability; Kyllonen et al., 1990). The Figure Matrices Test is part of the Cognitive Abilities Measurement test battery developed at Brooks Air Force Base. It is a 9-item computer-administered test of cognitive ability analogous to the paper-and-pencil Raven Advanced Progressive Matrices test (Raven, Raven, and Court, 1993). The scoring algorithm used for this test takes into account both the correctness of the test taker's answers and response times to generate the test score such that a higher score denotes better performance. Split-half and test-retest reliabilities of .68 and .59 respectively, have been reported by Kyllonen et al. (1990).

Spatial Processing Speed Test (Kyllonen et al., 1990). The Spatial Processing Speed Test is also part of the Cognitive Abilities Measurement test battery. It is a 12-item computer-administered test of spatial processing speed. Like the Figure Matrices Test, the scoring algorithm takes into account both the correctness of the test taker's answers and response times to generate the test score such that a higher score denotes better performance. Split-half and test-retest reliabilities of .70 and .42 respectively, have been reported by Kyllonen et al. (1990). An internal consistency of .72 was obtained for the present study.

Spatial Working Memory Test (Kyllonen et al., 1990). The Spatial Working Memory Test is also part of the Cognitive Abilities Measurement test battery. It is a 24-item computer-administered test of spatial working memory. Again, the scoring algorithm takes into account both the correctness of the test taker's answers and response times to generate the test score such that a higher score denotes better performance. Split-half and test-retest reliabilities of .86 and .69 respectively, have been reported by Kyllonen et al. (1990). An internal consistency of .82 was obtained for the present study.

Driving Behavior Questionnaire (Arthur and Doverspike, 1992). The dependent variables used in the present study, total crashes and number of moving violations, were obtained via a self-report driving behavior questionnaire. In completing the questionnaire, participants reported the total number of crashes in which they had been involved as one of the drivers. A crash was defined as any driving or traffic crash in which the participant was involved as one of the drivers, and in which a person had suffered physical injury (including fatalities) and/or there was $150 or more damage to property. Participants were also asked to report the total number of moving violation tickets they had received. Arthur and Graziano (1996) obtained a 2-3 day test-retest reliability of .98. The Driving Behavior Questionnaire was also used to collect information on the two demographic variables analyzed in the present study, namely age, and the total number of years the participant had been driving legally.

Procedure

As previously noted, participants were recruited to participate in a study on complex skill acquisition in which they were paid $5/hour for 16 hours of research participation. Each participant completed a computer-administered test battery that, in addition to a number of other individual difference measures, included the measures reported in this study (i.e., the Driving Behavior Questionnaire, Goldberg's Markers, the Computer-Administered Visual Attention Test, Figure Matrices Test, Spatial Processing Speed Test, and Spatial Working Memory Test). Instructions for completing the measures were given verbally and were also provided on the measures themselves. The measures were self-administered, and participants were tested in groups of four. The ordering of the tests was standardized as a lack of ordering effects had been previously shown (Arthur and Graziano, 1996).

Results

An examination of the crash and moving violations tickets data indicated that along with a low base-rate (mode = 0), they were non-normal (e.g., skewness = 2.83 and 2.15, respectively) with a high concentration of cases at the lower end of the distribution. This non-normality of crash data is potentially problematic with correlation coefficients (McGuire, 1973). Consequently, to correct for this problem, a log-linear transformation of the (non-zero) crash and ticket data was performed (Howell, 1992) resulting in a skewness of 1.92 and 1.02, respectively. Thus, with the exception of the descriptive statistics presented in Table 1, all analyses are based on the log-transformed data.

In terms of the zero-order correlations (see Table 1), both demographic/exposure variables (age and number-of-years-driving) were related to both crashes and tickets. Of the information processing variables, only spatial working memory was related to total crashes. None of the other variables were related to any of the criteria. Likewise, of the personality variables, only extraversion was related to crashes and tickets; none of the other personality variables were related to any of the criteria.

It is important to note that the primary objective of the study was not to focus on the validity of each individual variable but instead, to undertake a comparative and combined assessment of the three commonly used predictor *categories* of crash involvement, using one of them—demographic and exposure variables—as a control variable. To accomplish this, a series of hierarchical regression analyses were conducted with each category of variables being entered as a block. Demographic variables were entered first to permit an assessment of the effects of information processing and personality variables over and beyond these variables. Specifically, this procedure statistically controlled for the differences in age and years of driving experience. For information processing and personality variables, the two possible orders of entry were both tried and identical results were obtained for them. The results of the hierarchical analyses are presented in Table 2.

As the results in Table 2 indicate, the combination of the three predictor categories resulted in the ability to account for a much larger amount of variance for both criteria (i.e., 21% and 28% for crashes and tickets, respectively) than any predictor category by itself. Furthermore, for crashes, each predictor category predicted an equal amount of variance demonstrating that even when demographic and exposure variables were held constant, information processing and personality variables contributed significant incremental validity to the prediction of crash involvement. On the other hand, for tickets, demographic variables accounted for twice the amount of variance explained by personality variables (18% and 9%, respectively). However, the variance explained by information processing variables was negligible (1%)—a finding consistent with the observation that the sorts of activities that may result in one receiving a moving violation ticket are not ability-based, but instead, are typically volitional in nature (e.g., speeding, running a red light) and consist of the types of behaviors that one would expect to be associated with personality. Thus these results again demonstrate that when demographic and exposure variables were held constant, personality variables contributed significant incremental validity to the prediction of tickets.

Table 3 presents the standardized regression weights for each of the variables entered in the hierarchical analysis. For the personality variables, these results suggest that the variables important for the prediction of crashes (i.e., extraversion and agreeableness) may be different from those for tickets (i.e., extraversion, openness, and conscientiousness). Furthermore, a comparison of the results obtained for the zero-order correlations and the hierarchical regression suggests an enhancement (commonly referred to as "suppression") effect (McFatter, 1979) such that by using the totality of personality and information processing variables, variables that are not otherwise *directly* related to the criterion enhance or increase the predictive validity of other predictors (Conger, 1974; Lord and Novick, 1968; McFatter, 1979). For instance, in the present case, neither openness nor conscientiousness yield statistically significant zero-order correlations with number of tickets (see Table 1). However, when included in a hierarchical regression with agreeableness, extraversion, and emotional stability, they both have significant relationships with the criterion (see Table 3).

Discussion

The results obtained for Study 1 were supportive of the general propositions of the present research. Specifically, that because the totality of the individual is involved in driving and subsequent crash-related behaviors, it seems more appropriate to try to predict these behaviors using the gestalt of the individual (i.e., multiple predictors from multiple predictor categories) instead of separate facets of the person. However, in spite of the generally favorable results, there were certain characteristics of the study that served as potential limitations. Namely, the sample was a relatively small, all-male, relatively young sample that had been driving for a relatively short length of time (approximately 5 years). And, it is widely recognized that these are all variables that play an important role in influencing the likelihood of crash involvement. Consequently, Study 2 sought to constructively replicate and extend the results of Study 1 by using a relatively large and representative sample that completed a battery of tests that represented the same three predictor categories as Study 1 but used different specific measures of some of the categories.

Table 1. Descriptive Statistics and Correlations for Study 1

	CRITERIA		DEMOGRAPHIC		INFORMATION PROCESSING				PERSONALITY				
Variables	1	2	3	4	5	6	7	8	9	10	11	12	13
CRITERIA													
1. Crashes	---												
2. Tickets	.28**	---											
DEMOGRAPHIC													
3. Age	.22**	.38**	---										
4. Driving Years	.26**	.43**	.84**	---									
INFORMATION PROCESSING													
5. Cognitive Ability	-.01	-.02	-.13	-.07	---								
6. Spatial Processing Speed	.02	.01	.05	.06	.24**	---							
7. Spatial Working Memory	-.23**	-.02	-.01	.03	.32**	.37**	---						

Table 1. (Continued)

	CRITERIA		DEMOGRAPHIC		INFORMATION PROCESSING				PERSONALITY				
Variables	1	2	3	4	5	6	7	8	9	10	11	12	13
PERSONALITY													
8. Visual Attention[a]	.06	.06	-.04	-.06	-.23**	-.36**	-.53**	---					
9. Agreeableness	.14	.07	-.03	.03	.04	-.07	.01	.02	---				
10. Conscientiousness	.01	-.06	.08	.18*	.02	.06	.06	-.10	.47***	---			
11. Emotional Stability	-.01	.08	.06	.17*	.04	.16*	.09	-.07	.39**	.22**	---		
12. Extraversion	.18*	.17*	-.03	.08	-.06	-.01	-.03	-.03	.30**	.21**	.26**	---	
13. Openness	.02	-.10	-.06	-.01	.12	.10	.12	-.08	.26**	.16*	.12	.32**	---
M	1.02	1.94	20.41	4.97	86.59	31.98	43.19	23.31	137.24	126.06	83.31	115.63	138.65
SD	1.33	2.52	2.59	2.72	10.81	2.55	8.87	16.21	15.58	19.81	19.05	19.98	15.85
Minimum	0.00	0.00	15.00	1.00	58.30	21.44	23.81	2.00	94.00	60.00	52.00	70.00	95.00
Maximum	10.00	13.00	30.00	14.00	109.35	34.32	61.45	73.00	174.00	165.00	150.00	160.00	175.00

Note. Descriptive statistics are based on untransformed, raw data. Crashes and tickets were log-transformed before calculating their correlations. [a]Lower scores reflect better performance. * $p < .05$, ** $p < .01$, two-tailed. $N = 153$.

Table 2. Summary of Hierarchical Regression Results for Study 1

	Crashes	
Block	R^2	ΔR^2
1. Demographic	.07**	
2. Information processing	.14**	.07*
3. Personality	.21**	.07*
	Tickets	
Block	R^2	ΔR^2
1. Demographic	.18**	
2. Information processing	.19**	.01
3. Personality	.28**	.09**

Note. * $p < .05$,** $p < .01$. $N = 153$.

Table 3. Summary of Standardized Regression Weights for Study 1

	Crashes	
Variable	β	*t*
1. Age	-.01	0.07
2. Driving Years	.30	2.05*
3. Cognitive Ability	.08	0.94
4. Spatial Processing Speed	.14	1.60
5. Spatial Working Memory	-.30	3.19**
6. Visual Attention[a]	-.04	0.44
7. Agreeableness	.22	2.35**
8. Conscientiousness	-.14	1.58
9. Emotional Stability	-.16	1.85
10. Extraversion	.17	2.05*
11. Openness	-.04	0.53
	Tickets	
Variable	β	*t*
1. Age	.10	0.73
2. Driving Years	.37	2.63**
3. Cognitive Ability	.05	0.67
4. Spatial Processing Speed	.04	0.53
5. Spatial Working Memory	.03	0.39
6. Visual Attention[a]	.09	1.07
7. Agreeableness	.15	1.68
8. Conscientiousness	-.21	2.54*
9. Emotional Stability	-.04	0.52
10. Extraversion	.22	2.75**
11. Openness	-.17	2.22*

Note. [a]Lower scores reflect better performance. * $p < .05$, ** $p < .01$. $N = 153$

STUDY 2: METHOD

Participants

Participants were 333 paid volunteers recruited from Texas A&M University and its community by advertisements in the school and local newspapers and TV stations, as well as posted notices both on campus and in the community (e.g., libraries, grocery stores, and driver's license renewal offices). Individuals were recruited specifically for the present study and were required to have a valid Texas driver's license. The mean age of the study participants was 36.41 years (*SD* = 17.12) and the mean number of years they had been driving was 19.69 years (*SD* = 16.40). The sample was 57.36% female. Thirty-four percent of the participants were recruited from the psychology department's research participant pool and received course credit for participating in the study; the other participants were paid either $20 (28%) or $50 (38%). The payment was increased to $50 when it turned out to be quite difficult to recruit participants at the $20 rate.

Measures

The personality, and visual attention measures, and the driving behavior questionnaire were the same as those used in Study 1. However, three different information processing measures were used in Study 2 in addition to the CA-VAT. (A noteworthy difference in the CA-VAT was that at the time of the data collection for Study 2, the scoring algorithm had been modified such that test scores reflected the number of correct responses instead of the number of errors. Consequently, to be consistent with Study 1, the sign of the correlations for this measure were reversed in the presentation of the results.) The three different information processing measures—the Raven Advanced Progressive Matrices (cognitive ability), the Auditory Selection Attention Test (auditory attention), and the Group Embedded Figures Test (field dependence-independence)—are described below.

Raven Advanced Progressive Matrices (Raven et al., 1998). We used the short form of the Advanced Progressive Matrices (Arthur and Day, 1994; Arthur, Tubre, Paul, and Sanchez-Ku, 1999) as the measure of cognitive ability. The short form consists of a subset of 12 items from the 36 Advanced Progressive Matrices items which are arranged in an ascending order of difficulty. The Advanced Progressive Matrices is scored by summing the number of items correctly answered. We used an administration time of 30 minutes.

Auditory Selective Attention Test (Mihal and Barrett, 1976). In the Auditory Selective Attention Test, 24 dichotic messages are presented simultaneously to participants via stereo headphones. Each message consists of a pair of either single English letters or digits ranging from 0 to 9. A tone designates the target ear and the test taker is to report all numbers in the target ear. The score on the Auditory Selective Attention Test is the total number of errors (Doverspike, Cellar, and Barrett, 1986).

Group Embedded Figures Test (Witkin, Oltman, Raskin, and Karp, 1971). The Group Embedded Figures Test is a paper-and-pencil test of field dependence-independence. The Group Embedded Figures Test requires the test taker to perceptually identify a target geometric figure embedded within an irrelevant stimulus content. The score on the Group Embedded Figures Test is the total number of errors.

Table 4. Descriptive Statistics and Correlations for Study 2

	CRITERIA		DEMOGRAPHIC			INFORMATION PROCESSING				PERSONALITY				
Variables	1	2	3	4	5	6	7	8	9	10	11	12	13	14
CRITERIA														
1. Crashes	---													
2. Tickets	.14**	---												
DEMOGRAPHIC														
3. Age	.13**	-.05	---											
4. Sex[a]	.15**	.05	.00	---										
5. Driving Years	.14**	-.09	.97**	.01	---									
INFORMATION PROCESSING														
6. Cognitive Ability	-.02	.03	-.40**	.01	-.37**	---								
PERSONALITY														
7. Visual Attention[b]	.07	.04	.47**	-.08	.46**	-.56**	---							
8. Auditory Attention[c]	-.09	.00	.29**	.01	.27**	-.34**	.26**	---						
9. Field Dependence-Independence	.07	.00	-.39**	.13**	-.36**	.58**	-.49**	-.33**	---					
10. Agreeableness	-.14**	-.11*	-.11*	-.12*	-.11*	.04	-.08	-.06	.03	---				
11. Conscientiousness	-.17**	.10	.01	-.08	.00	-.02	-.05	.03	-.04	.38**	---			

Table 4. (Continued)

	CRITERIA		DEMOGRAPHIC			INFORMATION PROCESSING				PERSONALITY				
Variables	1	2	3	4	5	6	7	8	9	10	11	12	13	14
12. Emotional Stability	-.03	.10	.27**	.09	.26**	-.03	.08	.07	.03	.28**	.14**	---		
13. Extraversion	.08	.13*	-.08	-.02	-.06	.01	.03	-.04	.00	.20**	.09	.13**	---	
14. Openness	.07	.03	-.13**	.10*	-.11*	.29**	-.21**	-.18**	.22**	.24**	.16**	-.04	.19**	---
M	1.47	1.18	36.41	--	19.69	7.03	153.65	18.88	10.95	139.05	133.36	101.41	115.28	130.18
SD	1.71	1.65	17.12	--	16.40	3.07	57.97	22.71	5.55	15.45	17.89	20.66	20.71	19.45
Minimum	0.00	0.00	15.00	--	0.00	0.00	8.01	0.00	0.00	81.00	78.00	41.00	54.00	76.00
Maximum	11.00	12.00	85.00	--	70.00	12.00	306.29	205.00	18.00	117.00	174.00	165.00	166.00	178.00

Note. Descriptive statistics are based on untransformed, raw data. Crashes and tickets were log-transformed before calculating their correlations. [a]Female = 1, Male = 2. [b]To be consistent with Study 1, the sign of these correlations have been reversed such that lower scores reflect better performance. [c]Lower scores reflect better performance. * $p < .05$, ** $p < .01$, two-tailed. $N = 333$.

Procedure

Participants completed the battery of tests in individual sessions which lasted approximately two hours. As with Study 1, the ordering of the tests was standardized across all participants.

RESULTS

The results for Study 2, which are based on analyses analogous to those presented for Study 1, are shown in Tables 4-6. It should be noted that we again log-transformed the (non-zero) crash and ticket data to address the low base-rate and non-normality in these data. In terms of the zero-order correlations (see Table 4), all three demographic/exposure variables (age, sex, and number-of-years-driving) were related to crashes but not tickets. None of the information processing variables were related to either crashes or tickets. And for the personality variables, agreeableness and conscientiousness were related to crashes, and agreeableness and extraversion were related to tickets.

Table 5. Summary of Hierarchical Regression Results for Study 2

	Crashes	
Block	R^2	ΔR^2
1. Demographic	.05**	
2. Information processing	.06**	.01
3. Personality	.12**	.06**
	Tickets	
Block	R^2	ΔR^2
1. Demographic	.01	
2. Information processing	.02	.01
3. Personality	.10**	.08**

Note. ** $p < .01$. $N = 333$.

The results of the hierarchical analyses for Study 2 are presented in Table 5. Similar to Study 1, both demographic and personality variables explained a significant portion of the variance in crashes. In contrast to Study 1, information processing variables did not explain a significant portion of the variance in crashes. Also in contrast to Study 1, only personality explained a significant portion of the variance in tickets.Table 6 presents the standardized regression weights for each of the variables entered in the hierarchical analysis. A noteworthy result, similar to that obtained in Study 1 for tickets, is that combining predictors resulted in significant enhancement effects. This is apparent by comparing the correlations in Table 4 to the regression weights in Table 6. For crashes, extraversion did not yield a significant zero-order correlation but it did make a significant contribution in the regression model. Similarly, although neither conscientiousness nor emotional stability yielded statistically significant zero-order correlations with number of tickets, they both had significant relationships with tickets in the regression model. Hence this pattern of results for both crashes and tickets again shows and highlights the importance of taking a multivariate approach to the prediction of driving outcomes.

Table 6. Summary of Standardized Regression Weights for Study 2

	Crashes	
Variable	β	t
1. Age	.08	0.34
2. Sex[a]	.11	2.06*
3. Driving Years	.10	0.43
4. Cognitive Ability	-.07	0.99
5. Visual Attention[b]	-.02	0.25
6. Auditory Attention[b]	-.10	1.78
7. Field Dependence-Independence	.09	1.30
8. Agreeableness	-.12	1.99*
9. Conscientiousness	-.12	2.06*
10. Emotional Stability	-.06	0.95
11. Extraversion	.15	2.73**
12. Openness	.05	0.93
	Tickets	
Variable	β	t
1. Age	.18	0.72
2. Sex[a]	.01	0.08
3. Driving Years	-.33	1.37
4. Cognitive Ability	.00	0.01
5. Visual Attention[b]	-.04	0.60
6. Auditory Attention[b]	.05	0.81
7. Field Dependence-Independence	.02	0.34
8. Agreeableness	-.18	2.88**
9. Conscientiousness	-.13	2.22*
10. Emotional Stability	.17	2.72**
11. Extraversion	.16	2.86**
12. Openness	.10	1.75

Note. [a]Female = 1, Male = 2. [b]Lower scores reflect better performance. * $p < .05$, ** $p < .01$. $N = 333$.

DISCUSSION

As a constructive replication and extension of Study 1, Study 2 was designed to address some limitations with Study 1. This was accomplished by using a relatively large and representative sample that completed a battery of tests that represented the same three predictor categories as Study 1 but used, in some instances, different specific measures of these categories. As with Study 1, the results obtained for Study 2 were supportive of the need for taking a multivariate approach to the prediction of driving outcomes. This was especially highlighted by the enhancement effects found with the personality variables. In contrast to Study 1, information processing variables did not contribute to the prediction of crashes or tickets.

CONCLUSION

Relatively speaking, few studies have used multiple regression and other multivariate techniques in an attempt to isolate the unique contribution of particular variables to the prediction of driving outcomes (e.g., Arthur, Barrett, and Doverspike, 1990; Arthur and Doverspike, 1992; Hansen, 1989; Jacobs et al., 1996; Shope, Waller, and Lang , 1996). However, as noted by Elander et al. (1993), even these studies:

> "typically exclude other(s) [predictors] that have previously been shown to be associated with crash risk. They examine crash risk from a particular perspective, and this determines the choice of variables to be included. The fact that other variables already shown to be associated with crash risk might account for the relationships observed tends to be neglected." (p. 283).

The present research investigated not only the role of multiple predictors but also whether an empirical combination of three commonly (but independently) used predictor *categories* of crash involvement would result in incremental prediction over and above the single categories. Consistent with the distinction between errors and violations as anomalous behaviors that result in crashes, one would expect crash involvement to be influenced by both ability and personality factors since errors and violations have different underlying psychological basis corresponding to the commonly used predictor categories. Consequently, it was argued that a theoretical and empirical combination of multiple predictors from multiple predictor categories might be a more meaningful and fruitful approach to the prediction of driving outcomes. This premise was generally supported by our results. All three predictor categories made significant contributions to the prediction of driving outcomes in one of the four regression models (crashes in Study 2). In two of the four regression models (tickets in Study 1 and crashes in Study 2), both demographic and personality characteristics made significant contributions to the prediction of driving outcomes. Only in one model (tickets in Study 2) did a single predictor category (personality) make significant contributions.

Whereas the use of multiple predictor categories was not fully supported in all instances, taking a multivariate approach in general was strongly supported by our findings. Although the single predictor zero-order correlations were small, the amount of variance explained by combining predictors was substantially larger than that typically reported in the extant literature. For example, the amount of variance explained in crashes and tickets (21% and 28%, respectively) was substantially larger than the variance explained in Arthur et al.'s (1991) meta-analysis, in which the largest mean *r* was .26 (i.e., 6.76% variance explained) for selective attention. Furthermore, the results obtained in the present research also suggest the presence of an enhancement effect. An enhancement effect is an increase in the predictive validity of another variable (or set of variables) by the inclusion of other variables—enhancer variables—in a regression equation (Conger, 1974). "In effect, they [enhancer/suppressor variables] represent some aspect of the predictor variables that is not related to the criterion but functions to 'suppress' or subtract out this invalid component and thus to make the original predictor more valid" (Lord and Novick, 1968, pp. 271-272). The utility of enhancers, especially in the use of personality variables, has been further documented by Collins and Schmidt (1997). The potential role of enhancers lends additional support to the position that because the totality of the individual is involved in driving and subsequent crash- and moving violation-related behaviors, it seems more appropriate to try to predict these behaviors using the gestalt of the individual instead of single facets of the person. In other words, when one is driving, one does not leave one's "personality" at home and take along only one's information processing abilities or vice versa. Nor does one only take part of his or her personality with them. Therefore, when attempting to predict crash and moving violation involvement it would seem theoretically and conceptually more meaningful to examine the "whole" person instead single facets that might have limited predictive validity in the absence of the whole.

Although we did not find strong or consistent effects for information processing variables, which is not unusual (cf. Martín and Estévez, 2005), our results generally support the use of multiple predictor categories. In combining multiple predictor categories, the results of Study 1 suggest that high ability (information processing) individuals who have the "right" motivations, values, and attitudes and consequently make the right choices (personality variables) and also have lower exposure (demographic variables) have lower crash involvement levels than low ability individuals who make wrong choices and have a higher level of exposure. This approach is both theoretically and conceptually more meaningful and has empirical utility as well. Specifically, because the totality of the individual is involved in crash-related behaviors, it seems more appropriate to try to predict crash involvement using the gestalt of the individual.

Conceptualizing the prediction of crashes in this manner also provided some evidence suggesting that different psychological constructs may predict different elements of the driving behavior criterion space. Information processing variables were more likely to be related to crashes but not tickets—a finding consistent with the observation that the sorts of activities that may result in one receiving a moving violation ticket are not ability-based, but instead, are typically volitional in nature (e.g., speeding, running a red light) and consist of the types of behaviors that would be associated with personality variables. On the other hand, one would expect, both information processing and personality variables to be related to crash involvement (Hansen, 1989; Parker et al., 1995; Reason et al., 1990). However, the results across both studies presented here showed more consistent effects for personality, especially extraversion, than for information processing.

Interestingly, our results showing a relatively stronger relationship between personality and driving outcomes compared to information processing and driving outcomes mirrors the relatively greater level of attention that has been paid to personality as predictors of driving outcomes compared to the other two categories (e.g., see Table 1 of Arthur et al., 1991). We believe this attention to personality predictors is indeed worthwhile. In particular, research (e.g., Ulleberg and Rundmo, 2003) focusing on combining personality traits and testing causal models that include social cognitive variables as mediators is potentially promising as a means of furthering our understanding of these relationships.

In spite of the generally favorable results, there are a number of potential limitations that must be recognized and addressed in future research. First, our criterion data for crashes were limited to *total* crashes and were not disaggregated into fault and not-at-fault crashes. We chose not to disaggregate our data in this manner because doing so would have exacerbated the low baste rate and frequency issues that were present in our data. Furthermore, due to the methodological difficulties associated with collecting self-report data of this sort, it was not possible to collect these data at the disaggregated level of violations and errors (i.e., slips, lapses, mistakes, minor crashes, and major crashes). Consequently, it was not possible to map on specific crash-types to specific predictors and we only had indirect evidence suggesting that different psychological constructs may predict different elements of the driving behavior criterion space. Thus an assessment of the generalizability of our findings across these disaggregated levels of driving behavior criteria is warranted.

Second, although self-report accounts of crash involvement have been the most commonly used criterion (Ball and Owsley, 1991), asking drivers to report the motor vehicle crashes in which they have been involved opens up the possibility of either intentional or unintentional misrepresentations (see Elander et al., 1993; Harano, Peck, and McBride, 1975; Maycock, Lockwood, and Lester, 1991). For instance, when compared with state and insurance company records, it has been found that those older drivers with the highest number of recorded crashes tend to underreport crash involvement on self-report measures (Ball and Owsley, 1991). Furthermore, in a study cited by Loftus (1993), when 590 persons known to have been in injury-producing accidents during the previous year were interviewed, approximately 14% of them did not remember the accident a year later.

On the other hand, information acquired from state law enforcement agencies and insurance companies may also be flawed. The Insurance Institute of Highway Safety (1991) found that only 40% of motor vehicle crashes that should have been available to auto insurers appeared in publicly available records. Hakkert and Hauer (1988) also note that more crash victims are known to hospitals and insurance companies than to the police. Hence state and other archival records may be incomplete because they may underreport the number of actual crashes (Arthur et al., 2005; Burns and Wilde, 1995). Some reasons for this include (a) the parties involved chose not to report the crash, (b) the crash occurred in a different state, or (c) the crash was judged by the police to not be serious enough to report (McGuire, 1973; Smith, 1976). Lastly, mechanical and other nonbehavioral factors, such as the car engine catching fire, are often coded as accidents/crashes in archival records. State records can also be difficult to obtain, and the process can be time consuming and costly. So in addition to being able to canvass all crashes (Elander et al., 1993), from a practical standpoint, self-report information is much easier to collect. In summary, in concluding that self-report measures are not inherently inferior to archival data, Arthur et al. (2005) note that "self-report data appear to include a broader range of incidents (more crashes and tickets) and thus researchers should consider using self-report data when they wish to include lower threshold crashes and tickets that may be removed from a state record because of the completion of a defensive driving course" (p. 303).

Third, the reported findings should be replicated using longitudinal (predictive) designs instead of the retrospective (post-dictive) design used here. Although retrospective designs are more practical, feasible, and conveniently lend themselves to longer criterion time frames, because of their higher levels of methodological rigor, prospective designs allow the researcher to make stronger statements about relationships between predictors and crash involvement. They also permit an assessment of crash reliability with the periodic collection of crash data and the use of alternative criterion measures such as the time interval between crashes.

In conclusion, our findings contribute to a better understanding of the role individual differences play in driving outcomes and related subsequent efforts to encourage and increase safer driving behavior. Although it may not be readily apparent how enhancing the prediction of driving outcomes addresses safety issues, there are indeed several ways in which better prediction can be applied to promote safety and reduce the costs of motor vehicular crashes (Ulleberg and Rundmo, 2003).

From an employment standpoint, enhanced prediction of crash involvement can provide a means of better identifying individuals who may be less suitable for jobs that require extensive driving. In such a case, there are numerous potential costs such as collision and property damage, medical bills and workers' compensation, as well as psychological damage associated with trauma. With respect to the latter, it is even conceivable that a selection procedure that ignores the intrinsic human attributes required of safe driving practices may make an employer vulnerable to lawsuits concerning negligent hiring practices.

Enhanced prediction can, likewise, be utilized as a means of identifying youthful drivers who may need special (or additional) driver training. Such training may foster self-awareness, provide more salient and accurate depictions of the risks and hazards of driving, or concentrate more fully on the attention and behavioral components of safe driving practices. Similarly, enhanced prediction may increase an auto insurance carrier's ability to distinguish between high- and low-risk teens and young adults with respect to crash involvement. Thus, increased prediction may translate into more equitable insurance premiums. Ostensibly, high-risk youths would be discriminated against; however, an insurance carrier may offer those youths, deemed as a high-risk by a predictor test battery, some sort of special driver training similar to that previously mentioned as a means of offering a similar discount. Consequently, enhanced prediction can offer both more equitable insurance rates as well as promote safe driving practices. Such a use is analogous to situations where insurance carriers offer "good student" discounts to college students, but allow those who are college-aged and not attending college the opportunity to take a special driving class as a means of benefiting from a similar discount.

In summary, in spite of the limitations noted above, the results obtained for the present study generally suggest that consistent with the conceptual need to recognize the role of the "total" individual in driving-related behaviors, the use of multiple predictors from multiple predictor categories in the prediction of driving outcomes is empirically meaningful. As would be expected, this approach accounts for substantially more variance in the criteria than either any single predictor or predictor category by itself.

REFERENCES

Arthur, W. Jr., Barrett, G. V. & Alexander, R. A. (1991). Prediction of vehicular accident involvement: A meta-analysis. *Human Performance, 4,* 89-105.

Arthur, W. Jr., Barrett, G. V. & Doverspike, D. (1990). Validation of an information-processing-based test battery for the prediction of handling accidents among petroleum transport drivers. *Journal of Applied Psychology, 75,* 621-628.

Arthur, W. Jr., Bell, S. T., Edwards, B. D., Day, E. A., Tubre, T. C. & Tubre, A. H. (2005). Convergence of self-report and archival motor vehicle crash involvement data: A two-year longitudinal follow up. *Human Factors, 47,* 303-313.

Arthur, W. Jr. & Day, D. V. (1994). Development of a short form for the Raven Advanced Progressive Matrices Test. *Educational and Psychological Measurement, 54,* 394-403.

Arthur, W. Jr. & Doverspike, D. (1992). Locus of control and auditory selective attention as predictors of driving accident involvement: A comparative longitudinal investigation. *Journal of Safety Research, 23,* 73-80.

Arthur, W. Jr. & Doverspike, D. (2001). Predicting motor vehicle crash involvement from a personality measure and a driving knowledge test. *Journal of Prevention and Intervention in the Community, 22,* 35-42.

Arthur, W. Jr., Fuentes, R. & Doverspike, D. (1990). Relationships among personnel tests, age, and job performance. *Experimental Aging Research, 16,* 11-16.

Arthur, W. Jr. & Graziano, W. G. (1996). The five-factor model, conscientiousness, and driving accident involvement. *Journal of Personality, 64,* 593-618.

Arthur, W. Jr., Strong, M. H. & Williamson, J. (1994). Validation of a visual attention test as a predictor of driving accident involvement. *Journal of Occupational and Organizational Psychology, 67,* 173-182.

Arthur, W. Jr., Tubre, T. C., Paul, D. S. & Sanchez-Ku, M. L. (1999). College-sample psychometric and normative data on a short form of the Raven Advanced Progressive Matrices Test. *Journal of Psychoeducational Assessment, 17,* 354-361.

Ball, K. & Owsley, C. (1991). Identifying correlates of accident involvement for the older driver. *Human Factors, 33,* 583-595.

Ball, K., Sloane, M., Roenker, D. & Bruni, J. (1993). Visual attention problems as a predictor of vehicle accidents among older drivers. *Investigative Ophthalmology and Visual Science, 34,* 3110-3123.

Boggs, T. & Roberts, B. W. (2004). Conscientiousness and health-related behaviors: A meta-analysis of the leading behavioral contributors to mortality. *Psychological Bulletin, 130,* 887-919.

Burns, P. C. & Wilde, G. J. S. (1995). Risk taking in male taxi drivers: Relationships among personality, observational data and driver records. *Personality and Individual Differences, 18,* 267-278.

Butler, G. T. (1982). *Effectiveness and efficiency in driver education programs: Evaluation summary report 1966-81.* Traffic Safety Programs. Washington, DC: U. S. Department of Transportation, National Highway Traffic Safety Administration.

Clark, A. W. (1976). A social role approach to driver behavior. *Perceptual and Motor Skills, 42,* 325-326.

Collins, J. M., and Schmidt, F. L. (1997). Can suppressor variables enhance criterion-related validity in the personality domain? *Educational and Psychological Measurement, 57,* 924-936.

Conger, A. J. (1974). A revised definition of suppressor variables: A guide to their identification and interpretation. *Educational and Psychological Measurement, 34,* 35-46.

Craske, D. (1968). A study of the relationship between personality and accident history. *British Journal of Psychology, 41,* 399-404.

Dahlen, E. R. & White, R. P. (2006). The Big Five factors, sensation seeking, and driving anger in the prediction of unsafe driving. *Personality and Individual Differences, 41,* 903-915.

Digman, J. M. (1990). Personality structure: Emergence of the five-factor model. *Annual Review of Psychology, 41,* 417-440.

Doverspike, D., Cellar, D. & Barrett, G. V. (1986). The Auditory Selective Attention Test: A review of field and laboratory studies. *Educational and Psychological Measurement, 46,* 1095-1103.

Elander, J., West, R. & French, D. (1993). Behavioral correlates of individual differences in road-traffic crash risk: An examination of methods and findings. *Psychological Bulletin, 113,* 279-294.

Evans, L. (1991). Airbag effectiveness in preventing fatalities predicted according to type of crash, driver age, and blood alcohol concentration. *Accident Prevention and Analysis, 23,* 531-541.

Eysenck, H. J. & Eysenck, S. B. (1975). *Manual of the Eysenck Personality Questionnaire.* London: Hodder and Soughton.

Fine, B. J. (1963). Introversion-extraversion and motor vehicle driver behaviour. *Perceptual and Motor Skills, 12,* 95-100.

Goldberg, L. R. (1992). The development of markers for the Big-Five factor structure. *Psychological Assessment, 4,* 26-42.

Gopher, D. & Kahneman, D. (1971). Individual differences in attention and the prediction of flight criteria. *Perceptual and Motor Skills, 33,* 1335-1342.

Guastello, S. J. (1993). Do we really know how well our occupational accident prevention programs work? *Safety Science, 16,* 445-463.

Hakkert, S. & Hauer, E. (1988). The extent and implications of incomplete and inaccurate road accident reporting. In T. Rothengatter, and R. de Bruijn (Eds.). *Road user behavior: Theory and research.* Wolfeboro, NH: van Gorcum.

Hansen, C. P. (1989). A causal model of the relationship among accidents, biodata, personality, and cognitive factors. *Journal of Applied Psychology, 74,* 81-90.

Harano, R. M., Peck, R. C. & McBride, R. S. (1975). The prediction of accident liability through biographical data and psychometric tests. *Journal of Safety Research, 7,* 16-52.

Howell, D. C. (1992). *Statistical methods for psychology.* Boston, MA: Duxbury Press.

Insurance Institute for Highway Safety. (1990). *Status Reports, 24,* 7.

Insurance Institute for Highway Safety. (1991). *Status Reports, 26,* 1-10.

Jacobs, R. R., Conte, J. M., Day, D. V., Silva, J. M. & Harris, R. (1996). Selecting bus drivers: Multiple predictors, multiple perspectives on validity, and multiple estimates of utility. *Human Performance, 9,* 199-217.

Kyllonen, P. C., Christal, R. E., Woltz, D. J., Shute, V. J., Tirre, W. C. & Chaiken, S. (1990). *Cognitive Abilities Measurement (CAM) test battery* (Version 4.0) [Computer program]. Brooks Air Force Base, TX: Armstrong Laboratory.

Lajunen, T. (2001). Personality and accident liability: Are extraversion, neuroticism, and psychotocism related to traffic and occupational fatalities? *Personality and Individual Differences, 31,* 1365-1373.

Loo, R. (1979). Role of primary personality factors in the perception of traffic signs and driver violations and accidents. *Accident Prevention and Analysis, 17,* 449-460.

Lord, F. M. and Novick, M. R. (1968). *Statistical theories of mental test scores.* Reading, MA: Addison-Wesley.

Loftus, E. F. (1993). The reality of repressed memories. *American Psychologist, 48,* 518-537.

Martín, F. S. & Estévez, M. A. Q. (2005). Prevention of traffic accidents: The assessment of perceptual-motor alterations before obtaining a driving license. A longitudinal study of first years of driving. *Brain Injury, 19,* 189-196.

Massie, D. L., Campbell, K. L. & Williams, A. F. (1995). Traffic accident involvement rates by driver age and gender. *Accident Analysis and Prevention, 27,* 73-87.

Maycock, J., Lockwood, C. R. & Lester, J. F. (1991). *The accident liability of car drivers* (Rep. No. 315). Crowthorne, England: Transport and Road Research Laboratory.

McCrae, R. R. & Costa, P. (1987). Validation of the five-factor model of personality across instruments and observers. *Journal of Personality and Social Psychology, 52,* 81-90.

McCrae, R. R. & John, O. (1992). An introduction to the five-factor model and its application. *Journal of Personality, 60,* 175-216.

McEvoy, G. M. & Cascio, W. F. (1989). Cumulative evidence of the relationship between employee age and job performance. *Journal of Applied Psychology, 74,* 11-17.

McFatter, R. M. (1979). The use of structural equation models in interpreting regression equations including suppressor and enhancer variables. *Applied Psychological Measurement, 3,* 123-135.

McGuire, F. L. (1973). The nature of bias in official accident and violation records. *Journal of Applied Psychology, 57,* 300-305.

Mihal, W. & Barrett, G. V. (1976). Individual differences in perceptual information processing and their relation to automobile accident involvement. *Journal of Applied Psychology, 61,* 229-233.

National Highway Safety Administration (1989). *Conference on Research and Development Needed to Improve Safety and Mobility of Older Drivers* (Report DOT HS 807 554). Washington, DC: U.S. Department of Transportation.

National Safety Council (1997). *Accident Facts, 1997 Edition.* Itasca, IL: Author.

Parker, D., Manstead, A. S., Stradling, S. G. & Reason, J. T. (1992). Intention to commit driving violations: An application of the theory of planned behavior. *Journal of Applied Psychology, 77,* 94-101.

Parker, D., Reason, J. T., Manstead, A. S. & Stradling, S. G. (1995). Driving errors, driving violations and accident involvement. *Ergonomics, 38,* 1036-1048.

Parker, D., Stradling, S. G. & Manstead, A. S. (1996). Modifying beliefs and attitudes to exceeding the speed limit: An intervention study based on the theory of planned behavior. *Journal of Applied Social Psychology, 26,* 1-19.

Raven, J., Raven, J. C. & Court, J. H.(1993). *A manual for Raven's Progressive Matrices and Vocabulary Scales.* London: H. K. Lewis Co.

Reason, J. (1990). *Human error.* New York: Cambridge University Press.

Reason, J., Manstead, A., Stradling, S., Baxter, J. & Campbell, K. (1990). Errors and violations on the roads: A real distinction? *Ergonomics, 33,* 1315-1332.

Robinson, B. S. & Bownas, D. A. (1993). *Predicting job performance of rail traffic controllers with an auditory selective attention test.* Paper presented at the 8th Annual Conference of the Society of Industrial and Organizational Psychology, San Francisco, CA.

Schuster, D. H. & Guilford, J. S. (1964). The psychometric prediction of problem drivers. *Human Factors, 6,* 393-421.

Shope, J. T., Waller, P. F. & Lang, S. W. (1996). Alcohol-related predictors of adolescent driving: Gender differences in crashes and offenses. *Accident Analysis and Prevention, 28,* 755-764.

Singh, A. P. (1978). Neuroticism, extraversion and accidents. *India Psychological Review, 16,* 41-45.

Smith, D. I. (1976). Official driver records and self-reports as sources of accident and conviction data for research purposes. *Accident Prevention and Analysis, 8,* 207-211.

Strong, M. H. (1992). *An assessment of the criterion-related validity and improvement in the utility of the Computer-Administered Visual Selective Attention Test.* Unpublished Master's thesis. Texas AandM University.

Ulleberg, P. & Rundmo, T. (2003). Personality, attitudes and risk perception as predictors of risky driving behaviour among young drivers. *Safety Science, 41,* 427-443.

Wilson, R. & Greensmith, J. (1983). Multivariate analysis of the relationship between drivometer variables and drivers' accident, sex and exposure status. *Human Factors, 25,* 303-312.

Witkin, H. A., Oltman, R. K., Raskin, E. & Karp, S. A. (1971). *A manual for the Embedded Figures Test.* Palo Alto, CA: Consulting Psychologists Press.

In: Accidents: Causes, Analysis and Prevention
Editors: Henri Bédard and Geraud Delashmit
ISBN: 978-1-60741-712-5

Chapter 9

THE ROLE OF SITUATION AWARENESS IN ACCIDENT PREVENTION

***Steven J. Kass*[1], *Kerstan Cole*[2] *and Shauna Legan*[3]**
[1]University of West Florida, Pensacola, FL, USA
[2]Texas Tech University, Lubbock, TX, USA
[3]University of West Florida, Pensacola, FL, USA

ABSTRACT

This chapter reviews the literature on distracted driving and makes the case that the focus of the research and legislation designed to prevent accidents is often misdirected. For instance, as of now (or in the near future) the Insurance Institute for Highway Safety (2007) noted that only six U.S. states have bans on handheld cellular phone use among drivers, but none of these states ban the use of hands-free cellular phones. These laws appear to address accident prevention by ensuring that drivers have both hands free to operate their vehicles. However, recent studies have shown that visual and cognitive distractions (rather than occupied hands) are the main contributing factors to traffic accidents in which cellular phones were currently in use. In fact, several studies have shown that impaired driving performance is just as likely to occur whether the phone is held or hands-free (e.g., Horrey & Wickens, 2006). The current chapter argues that cellular phones (and other nonessential devices) contribute to traffic accidents by impairing drivers' situation awareness (SA). That is, conversing with an individual who is not in the car, and who is therefore unable to alter the conversation based on current driving conditions, impairs drivers' ability to identify, attend to, and quickly react to relevant environmental stimuli and situations. Though situation awareness has been studied extensively in military pilots, this research is only recently being applied to the population of drivers. The relevance of SA to safe driving and potential methods for maintaining SA are addressed. These methods include technological advances such as tactile cueing and warning systems, as well as often overlooked psychological approaches for improving concentration.

INTRODUCTION

In July of 2007, a highly publicized accident in upstate New York occurred in which a 17 yr old driver of a sport utility vehicle crashed just after she passed another car and swerved back into oncoming traffic slamming into a tractor trailer (Associated Press, July 15, 2007). She and the other 4 teenagers on board were all killed. Authorities confirmed that her cellular phone was in use, verifying that calls and text messages were sent right at the time leading up to the accident. While it cannot be confirmed whether the driver was actually the one sending or receiving the calls/text messages, the cellular phone was believed to be a contributing factor to the crash. Similar stories have appeared all too frequently in newspapers across the country. Given the number of cellular phone users on the road, this should not be surprising. According to recent data provided by the Cellular Telecommunications and Internet Association (CTIA; 2007) more than 75 percent of Americans (i.e., 233 million) subscribe to cellular phones services. In a survey conducted by a large insurance company, 73 percent of the 1200 respondents reported that they held conversations on their cellular phones while they were driving (Insurance Institute for Highway Safety, 2007). From naturalistic observations of drivers idle at stop signs and stoplights (during daylight hours), the National Highway Traffic Safety Administration (NHTSA) and the National Center for Statistics and Analysis (Glassbrenner & Ye, 2007) estimated that 5 percent of drivers (745,000 vehicles) on the road at any one time are using hand-held cellular phones (estimate did not include drivers using headsets). Though this percentage is down slightly from the previous year's observation, cellular phone conversations and text messaging comprise just some of the many distractions posing a threat to drivers (e.g., navigation systems, food and beverages, drugs and alcohol). Although motorists may trivialize the risks associated with driver distraction, other reports have further indicated the extent of the problem. For example, Virginia Tech Transportation Institute, in collaboration with NHTSA (Klauer, Dingus, Neale, Sudweeks & Ramsey, 2006), found that driver inattention was involved in 80 percent of all vehicle crashes (65 percent of near-crashes). The use of cellular phones was the most frequently reported category of distraction (i.e., more than eating, grooming, talking to passengers, etc.). According to data from the Fatality Reporting System (FARS) in 1999, driver distraction was a contributing factor in 11% of fatal crashes involving 4,462 fatalities (Glassbrenner, 2005). Additionally, between 25 and 30% of injury and property damage cases were attributed to driver distraction (Utter, 2001).

The controversy generated by the public's use of cellular phones while driving has impacted some lawmakers who seek to make roadways safer. As of yet, there is little empirical evidence as to whether laws prohibiting cell phone use have actually impacted accident statistics. As of this writing, only five states (California, Connecticut, New York, New Jersey, and Washington) and the District of Columbia have jurisdiction-wide bans on the use of cellular phones while driving (IIHS, 2007). Additional states allow localities to determine bans on cellular phone use and other states have bans on the use of cellular phones by school bus drivers and young drivers. Surprisingly, eight states specifically prohibit localities from banning cellular phone use by drivers and none of the 50 states prohibit the use of hands-free cellular phones while driving. Though the number of states that are at least considering cellular phone restrictions among drivers is on the rise, the fact that no state has of yet banned the use of hands-free cellular phones suggests that the growing body of empirical research on the topic has been largely ignored by law makers. That is, numerous studies have demonstrated that the phone medium (hands-free or hand-held) matters much less than does the impact of the conversation on driving (e.g., Consiglio, Driscoll, Witte, & Berg, 2003; Redelmeier & Tibshirani, 1997). Consiglio et al. (2003) strongly argued that law makers' assumptions underlying the ban on hand-held cellular phones, but not hands-free phones, are mistaken.

Over the past 10 to 15 years, researchers (e.g., Beede & Kass, 2006; Garcia-Larrea et al., 2001; McKnight & McKnight, 1993; Recarte & Nunes, 2003) have provided strong empirical evidence that many aspects of one's driving performance suffer as a result of cognitively distracting conversations, especially those that involve spatially oriented information, are emotionally charged, or are otherwise engaging. For example, Horrey and Wickens (2006) concluded that cellular phones adversely impact driving especially in terms of slowed response times (e.g., to road hazards), though the data are equivocal regarding the impact on lane maintenance behavior (i.e., compensatory tracking). The authors suggested that this may be because lane maintenance is more of an automatic driving task requiring few cognitive resources, whereas tasks such as responding to hazards are more deliberate and occur later in the information processing sequence (i.e., decision making and response selection). Additionally, they found that laboratory experiments may actually underestimate the true effect of cellular phones as even greater impacts were found in field studies.

Impact of Hands-Free Cellular Phones on Driving

While hand-held cellular phones are dangerous distracters, hands-free cellular phones are well-documented to impair driving performance (Beede & Kass, 2006; Patten, Kircher, Ostlund, & Nilsson, 2004; Strayer & Johnston, 2001; Treffner & Barret, 2004). Motorists may assume that because their hands are free to control the car, they are able to maintain conversations on a cellular phone without experiencing impaired driving performance. The vast majority of empirical research has suggested that this assumption is false (Horrey & Wickens, 2006). Cognitive distractions such as the act of talking on a cellular phone while driving are caused by an internal cognitive context and are usually elicited by one's own thoughts or conversations with other people. Holding the phone in one's hand merely exacerbates the problem in much the same way as holding a beverage or reaching for an object may interfere with manual control of the vehicle. Evidence of the impairment caused by cellular phone conversation has been demonstrated by the many studies that have examined hands-free cellular phones in order to control for the effects holding a phone has on manual control of a vehicle. For example, in a simulator study, Beede and Kass (2006) found that the use of a hands-free cellular phone led to more traffic violations (e.g., running stop signs), greater attention lapses (e.g., stopping at green traffic lights), and longer delays at stop signs. These results were indicative of drivers' loss of attention; clearly not the loss of vehicle control. Furthermore, Consiglio et al. (2003) tested the braking response times of drivers in a simulated task that did not require any hand control of a vehicle (i.e., no steering). By using a braking task they were able to remove any effects that physically holding a cellular phone would have on a crucial driving task, thereby testing the structural interference hypothesis. That is, the primary task required the physical resources of visual fixation (watching a simulated brake light) and a foot-controlled response (stepping on brake), whereas the secondary conversation task required cognitive resources, and the use of one's hand only in the hand-held cellular phone condition. Participants displayed equally delayed responses in the hand-held and hands-free phone conditions despite the fact that the primary and secondary tasks did not compete for physical resources. Treffner and Barrett (2004) found similar effects of using a hands-free phone in a real-driving task. They suggested that sub-optimal braking behaviors were caused by drivers' impaired awareness and perception of the road environment when conversing. Such delayed braking behavior in traffic could cause a driver to have less time for decision making and response selection, making accidents more difficult to avoid

While more and more research documents a decrement in driving performance across dual-task cellular phone conditions, people continue to hold illusory beliefs about their own driving performance and their ability to operate automobiles during divided attention conditions. The increasing use of multi-functional, high-tech gadgets in cars (e.g., palm sized computers, or even in-dash DVD players) is one example of how people perceive that they are capable of multi-tasking while driving. In fact, Horswill, Waylen, and Tofield (2004) surveyed 181 drivers about their hazard perception, vehicle control skills, or general driving skill. Drivers rated their own driving skills as superior to that of their peers and the average driver on 18 out of 52 components of driving skill. The hazard-perception skill category accounted for the majority of the bias as drivers rated this skill as far superior to other driving skills such as vehicle control skills or general driving skills. Horswill et al. suggested that drivers' illusory beliefs about their hazard-perception skills may actually affect their own safety

To many drivers, the impact of using a cellular phone may not be readily apparent because people regularly hold conversations with passengers while successfully maintaining their driving performance. However, unlike callers on the other end of a cellular phone, due to their own awareness of the environment passengers are able to convey various cues to help drivers focus their attention on the road. For example, if a car suddenly swerved in front of a driver, the passenger might stop talking altogether or provide a verbal warning or hand gesture to the driver. Conversely, drivers who are recipients of cellular phone calls do not receive continuous cues because physical location separates them from the caller. Because of this, conversations using cellular phones are qualitatively different, and potentially more dangerous, than conversations with in-vehicle passengers (e.g., Drews, Pasupathi, & Strayer, 2004; Strayer & Drews, 2007). Drews, Pasupathi, and Strayer (2004) noted that passengers conversing with the driver were more likely to make references to traffic conditions and thereby aid, rather than distract, the driver through their shared awareness of the environment. This was demonstrated in that drivers who engaged in conversation with passengers were less likely to commit errors in a simulated driving task (e.g., miss an exit) than were drivers engaged in cellular phone conversations. On the other hand, Horrey and Wickens' (2006) meta-analysis showed little difference between the effects of cellular phone and passenger conversations on driving. However, they cautioned that these findings were based on very few studies that actually examined the effects of passenger conversations.

SITUATION AWARENESS AND DRIVING

Situation Awareness Defined

Perhaps the best way to explain the impact of cellular phones on driver safety is that these cognitive distractions disrupt drivers' ability to maintain their full awareness of driving-relevant cues. That is, drivers conversing on phones lack *situation awareness* (SA). Though the literature on SA dates back to mostly military research of the 1980's, it has recently gained popularity as an area of study in other dynamic, complex environments (e.g., medicine, sports, driving). SA can be defined as "the perception of the elements in the environment within a volume of time and space, the comprehension of their meaning and the projection of their status in the near future" (Endsley, 1988, p.97). Endsley conceptualized SA as comprised of the following three levels: perception (level 1), comprehension (level 2), and projection (level 3). Perception involves selective attention and is a necessary condition for higher levels of SA to occur. In fact, it forms the foundation on which these higher levels are built. Level 1 SA is achieved through perception of relevant cues in the environment and allows motorists to achieve a mental model or representation of the situation. For example, drivers rely on level 1 SA when they identify visual cues from the roadway such as when the brake lights illuminate on the car they are trailing or a foreign object appears on the road. Ma and Kaber (2006) suggested that level 1 SA is necessary for such operational tasks as steering and braking.

Once a driver has perceived the fragmentary elements of level 1 SA, these elements are assembled through the process of interpretation and comprehension, which involve the use of both working and long term memory. After accurately identifying task-relevant features of the environment, drivers may be able to recognize stimulus patterns during level 2 assessment. That is, drivers identify the pattern of stimuli (e.g., traffic lines, surrounding vehicles, current speed) and understand its meaning in order to quickly determine safe maneuvering behaviors (e.g., lane changes). The ability to learn patterns of information is believed to be a key component for developing situation awareness (Durso & Gronlund, 1999; Kass, Herschler, & Companion, 1991) and is often what distinguishes expert from novice performance in a variety of tasks (Chase & Simon, 1973). The highest level of SA (level 3) represents drivers' abilities to make predictions regarding the future status of the cues interpreted during level 2 SA. For example, drivers must estimate the spacing needed and speed of oncoming traffic in order to safely pass another vehicle. Level 3 SA signifies operators' highest levels of comprehension and their ability to anticipate future events. Projection of near future status is necessary for quality decision making and response selection in order to prevent collisions. Although SA is affected by various aspects of human information processing (e.g., cognition, working memory, attention, and workload), it is not equated with this process. Likewise, a loss of SA affects decision-making and performance respectively, but each of these is considered a separate construct (Endsley, 1995).

Research on SA and Driving

Though the research on the impact of SA on driving is rather sparse, what is available clearly demonstrates the importance of this research topic. For instance, Gugerty (1997) examined how drivers obtained necessary environmental knowledge (e.g., location of vehicles) while driving; a crucial component of attaining SA and safe vehicle operation. He proposed that drivers rely on both explicit and implicit knowledge. Explicit knowledge may be thought of as active knowledge in working memory (e.g., identifying the deceleration of a lead vehicle), whereas implicit knowledge is less active knowledge relevant to the current situation, but still accessible from long term memory (e.g., knowledge that a car may be in one's blind spot because the driver had passed that car earlier). Gugerty found that drivers (in a simulator study) relied much more heavily on explicit knowledge in maintaining their awareness of the situation despite the fact that information from implicit knowledge may, at times, be more crucial. For example, drivers focused more of their attention on cars in their immediate visual field, while ignoring cars in their blind spot that may pose the greatest risk. He theorized that this was likely due to the greater working memory demand of having to track cars in the blind spot. Though Gugerty did not study distractions caused by a secondary task such as cellular phone use, one can extend these results to the literature on distracted driving. That is, if increased working memory demands are placed on drivers by having to monitor their blind spots, then any additional working memory load (i.e., cellular phone conversation) ought to exacerbate the problem. In fact, in a second experiment in the same article, Gugerty found that at higher memory loads, drivers were less able to avoid crashes despite their attempts to focus attention on vehicles that presented the greatest potential hazard.

Kass, Cole, and Stanny (2007) examined the impact of cellular phone use on both experienced and novice drivers' SA. Using a variation on Endsley's (1995) Situation Awareness Global Assessment Technique (SAGAT), situation awareness was measured by periodically stopping the driving simulator and asking participants relevant questions about the visual scene just encountered (level 1 and 2 SA). Not surprisingly, drivers who were engaged in a cellular phone conversation answered significantly fewer questions correctly than did the control group. Perhaps more surprising was the finding that experienced drivers showed the same decrease in SA as a result of the cellular phone as did the novices (a group of high school teenagers enrolled in driver education). That is, having greater driving experience did not prevent the loss of SA. In addition, the same pattern of results was found for a variety of performance measures (i.e., various traffic violations) and an ecologically-based measure of level 3 SA (i.e., number of missed turns on a route following task). Overall, the results indicated that the SA and performance of experienced drivers engaged in cellular phone conversation was about the same as that of young (non-conversing) individuals who had never driven before.

Cognitive distractions become particularly important safety issues when motorists are navigating through changing traffic patterns. Under such distracting circumstances, attention and hazard detection, aspects of SA, are adversely affected (Strayer & Johnston, 2001). Garcia-Larrea, Perchet, Perrin, and Amendedo (2001) demonstrated, using event-related brain potentials (ERPs), that study participants' who were engaged in a signal detection task decreased their attentional allocation to sensory stimuli [level 1 SA] during phone conversation whether the phone was hand-held or hands-free. They concluded that two separate psychophysiological mechanisms were responsible for participants' degraded reaction time performance. First, the increased cognitive workload of carrying out a secondary task such as a phone conversation decreased participants' attention to sensory inputs (regardless of whether phone was hands-free). The second mechanism they identified is that the conversation reduced the participant's preparedness to make a motor response. Not surprisingly, the impact from the second mechanism was even stronger in the hand-held phone condition.

Other evidence has suggested that the brain has ways of compensating for a lack of attentional resources (e.g., those needed for driving) when faced with a secondary task (e.g., a phone conversation). Langer, Holzner, Magnet, and Kopp (2005) demonstrated that the cognitive system compensates for a lack of cognitive resources by tunneling attention on the central field of vision while ignoring peripheral visual stimuli [level 1 SA], thus limiting the amount of information available to form a complete mental model of the situation [level 2 SA]. They found that this reduction in peripheral visual attention, as measured by slowed reaction times to stimuli, was similar to that which occurs when individuals are under the influence of alcohol. Likewise, Strayer, Drews, and Crouch (2006) found in a simulator study that while the impact of cellular phone use on drivers' attention and impairment of performance may be qualitatively different from that of drivers under the influence of alcohol (at the legal limit of intoxication), the impairment was equally great. In fact, drivers in the cellular phone condition had slower braking reaction times, took longer to regain speed, and were involved in more accidents than were drunk drivers.

How Cellular Phones Affect SA

Cognition and Working Memory. Unlike perceptual processes that exist at level 1 SA, comprehension requires greater mental effort, time, and attention, and is guided by working memory. Driving-related tasks such as the comprehension of maps and directions and planning city routes through traffic are both processes that rely on cognition and working memory (Foo, Warren, & Duchon, 2005). The main component of working memory, the central executive, is generally referred to as "attention." According to Baddeley and Hitch (1974), the central executive has a limited capacity and is used during cognitively demanding tasks. For example, drivers following a set of directions must allocate attentional resources to both the directions and the task of driving through a changing environment.

Working memory processes those tasks with the highest demands first. Consequently, those tasks that are cognitively demanding, such as talking on a cellular phone and driving, require more resources from the central executive. Further, in a revised version of his first working memory model, Baddeley (1986) proposed that a visuo-spatial sketchpad and an articulatory loop serve the central executive. The visuo-spatial sketchpad provides a representation of external spatial imagery in one's mind. The external spatial stimuli represented by the sketchpad and the imagery elicited by cellular-phone conversations may conflict, causing either driving performance to suffer or the conversation to deteriorate. For example, because of its spatial nature, answering a geometry question is more likely to interfere with driving than is answering a question of a declarative nature, such as what is your favorite book?

Conversely, the articulatory loop serves to direct speech-based information through auditory presentation, such as through sub-vocal articulation (Eysenck & Keane, 1990). In regard to SA, the articulatory loop represents inner speech about progress and decision-making in a particular state. The expression of drivers' progress, through the articulatory loop, may interfere with their ability to maintain SA because the central executive does not allow for equal attentional resources for each of these processes. Imagine a situation in which you are trying to keep a telephone number in working memory (rote rehearsal) as you are trying to decide the correct street on which to turn in an unfamiliar city.

Divided Attention. Dual task situations require people to allocate various attentional resources. Breakdowns in performance occur when attentional resources are not proportionally distributed among tasks. This means that when the total demand of these tasks is excessive, one task or the other must suffer (Eysenck & Keane, 1990). Drivers are often faced with dual and multiple task situations. For example, they must be able to scan the environment for possible hazards while maintaining a safe speed and distance from other vehicles, perhaps while also attending to child passengers riding in the back seat. Current research utilizes theories of divided attention, rather than selective attention, due to the inherent conscious nature of dual-task performance (Haigney, Taylor, & Westerman, 2000; Mathews, Legg, & Charleton, 2003; Strayer, Drews, & Johnston, 2003).

Motorists who engage in cellular phone conversations are exercising a conscious choice to entertain their conversations while attempting to navigate through road conditions. Ultimately, they are attempting to divide their attention. Conversely, selective attention occurs when people focus on a subset of information while ignoring other variables. The goal of multi-tasking is to allocate resources proportionately across tasks without disrupting performance. The tasks with the highest processing priority will be processed first. In some cases, experts can perfect-time share between two tasks when performance of one task reaches automaticity. You may have experienced a driving situation where you reached your destination without being able to recall much about the travel; seemingly through autopilot. However, in more demanding traffic conditions, driving often requires deliberate actions and greater awareness, and therefore it is unlikely that performance ever fully reaches the level of automaticity

Recent technological advances provide more opportunities for drivers to experience multitasking. Despite the potential hazards, people continue to use their cellular phones while they drive. In one divided attention study, Recarte and Nunes (2002) posited that high mental load may cause drivers to lose control of their speed. In this study, conducted in a real traffic environment, participants were asked to perform secondary tasks while driving, such as either producing or shadowing a list of letters. Producing a list of letters is analogous to conversing on a cell phone because the conversation is self-generated. The authors found that the cognitive act of producing a list of letters, as compared to shadowing, interfered with the visual image of the road, thereby decreasing performance. One interpretation of the results is that the demands of the production tasks may have exhausted the capacity of the central executive because they were more demanding than the driving task. A driver's inability to diagnose a situation increases the probability that an error will occur regardless of whether this occurs during perception (level 1 SA) or comprehension (level 2 SA). One practical application of this study is that intense cellular phone conversations that require continuous mental representation and inner speech may conflict with one's ability to maintain SA.

Inattention Blindness. Strayer et al. (2003) investigated the effects of hands-free cellular phone conversations during simulated driving conditions. In this experiment, participants drove a simulator and followed a pacer vehicle that would stop at irregular intervals. They found that the conversations impaired drivers' reactions to cars braking in front of them and also impaired explicit recognition for roadside billboards. After reviewing eye-tracking data, the authors suggested that the cellular phone conversations impaired implicit memory for items presented at fixation. These results support the evidence that cellular phone conversations actually decrease the amount of visual input that the brain processes thereby impairing perception (level 1 SA) and performance. This phenomenon of looking without seeing has been coined "inattention blindness." Researchers who study inattention blindness posit that attention is a necessary condition for the conscious perception of any object (Ellis, 2001; Mack, 2003; Mack, Pappas, Silverman, & Gay, 2002). Strayer and Drews (2007) reported on four of their studies that, in the aggregate, tested the theory of inattention blindness as an explanation for the adverse impact of cellular phones on driving. The Strayer and Drews studies suggested that one task clearly *does* interfere with the other despite the apparent lack of overlap in resources required of each. They suggested that perhaps this was due to the phone task causing a bottleneck in the processing of information from the driving environment, thus forcing information from the two distinct tasks to be processed serially, rather than in parallel. This is consistent with the previously reviewed literature that suggested more demanding tasks are processed first. The Strayer and Drews (2007) studies present a challenge to Wickens' (1984) multiple resources model that suggests that an auditory task requiring a verbal response (e.g., phone conversation) could be performed concurrently with a visual/spatial task requiring a manual response (e.g., driving) with little interference

The nature of the interference caused by cellular phones has been the focus of other experiments involving inattention blindness. In one such study, McKnight and McKnight (1993) exposed participants to a 25-minute driving sequence containing 45 traffic situations and were expected to respond by manipulating controls of a simulated vehicle. Participants were also exposed to five distracter conditions that consisted of placing a cellular phone call, having a casual conversation, having an intense cellular phone conversation, tuning a radio, and a no distracter condition. All of the distracter conditions led to an increased failure to respond to the simulated traffic conditions. Essentially, the participants of this study were *looking* without *seeing* (Posner, 1980), or not perceiving the appropriate cues

People regularly engage in selective looking tasks during their daily commutes, such as detecting and avoiding dangerous situations as they navigate through traffic. However, when they are engaged in cellular phone conversations, their ability to detect hazardous situations is negatively affected because they are engaged in a cognitive context other than the road (Strayer & Johnston, 2001). These cognitive distractions cause cognitive tunneling that may result in a loss of awareness of peripheral objects (Peters & Peters, 2002). This provides one explanation for the results of the aforementioned McKnight and McKnight (1993) study. Relative to the control condition, the distracter conditions increased the participants' cognitive load. This was especially true during the intense cellular phone conversation. In real-life scenarios, as drivers become deeply involved in conversations, their cognitive load may increase perhaps causing restricted peripheral vision (Langer et al., 2005), limited field of view, or complete loss of SA (Peters & Peters, 2002)

Improving Situation Awareness

Technological Advances

Because SA is now gaining recognition as a major component of safe driving, some recent research has focused on advances to improve SA in both vehicle design and driver training. Within vehicle design, warning systems are becoming more prominent including haptic (tactile feedback) displays (Fitch, Kiefer, Hankey, & Kliener, 2007), lane departure warnings (Kozak et al., 2006), and collision avoidance warnings (May, Baldwin, & Parasuraman, 2006). Although these advanced warning systems can bring attention to possible hazards, drivers must be able to comprehend the warnings and apply appropriate actions.

Ma and Kaber (2006) found that a simple conversational task, such as responding to a telemarketing survey, negatively impacted SA. However, they found that using a 100% reliable navigation aid (as compared to less reliable aids) improved both level 2 and level 3 SA as assessed via SAGAT. Oddly however, the same system had an adverse effect on level 1 SA. They speculated that this counter-intuitive result occurred because the navigation aid may have made drivers complacent in terms of searching their environment for appropriate cues. Manser, Ward, Kuge, and Boer (2004) evaluated a haptic driver support system (DSS) for improving SA. The DSS provided continuous force feedback through the accelerator pedal to convey information about the gap (and closing rate) between the driver and a lead vehicle. As the criticality of the situation increased, the force on the driver's foot increased proportionately. Distracted drivers (performing a task requiring the manipulation of a music device) responded more quickly and appropriately when they were provided feedback from the DSS. Such a device could have great implications for reducing rear end collisions.

Other tactile feedback displays have shown promise as well (e.g., Diamond, Kass, Andrasik, Raj, & Rupert, 2002; Hopp, Smith, Clegg, & Heggestad, 2005; Raj, Kass, & Perry, 2000). For example, the Tactile Situation Awareness System (TSAS), consisting of a set of vibrating nodes like those found in a pager, may be used to provide vibro-tactile cues to drivers. Raj et al. (2000) used the TSAS system to provide feedback to participants in a simulated helicopter hovering task. The system (worn as a vest) provided vibratory feedback to study participants (i.e., active duty military pilots) indicating velocity and direction of drift from the target point. The pilots were also engaged in a cognitively demanding secondary arithmetic task which also drew their vision away from the simulated scene. Pilots maintained a more constant hover (i.e., less deviation from target) when they performed the task in the TSAS condition. However, though performance improved, a retrospective self-reported measure of SA showed no difference between conditions. This finding may be more a function of the way SA was assessed than any true lack of SA differences, or as Wickens, Lee, Liu, and Gordon-Becker (2004) noted, self –report measures of SA suffer in that "people are not always aware of what they are not aware" (p. 144). Diamond et al. (2002) demonstrated that TSAS was effective in alerting and directing pilots toward cockpit gauges that indicated potential aircraft problems (e.g., low fuel level). TSAS improved pilots' accurate identification of problematic gauges and response times. Perhaps such a system may be adapted to use in ground vehicles to help drivers identify when nearby vehicles have driven into the blind spot, thus preventing side collisions. The benefit of tactile cueing over other alerting mechanisms (assuming Wickens' cognitive resources model is correct) is that the tactile medium is greatly underutilized providing less competition for information conveyed via this modality. That is, while a driver may be visually distracted (e.g., by reading a text message), aurally distracted (e.g., in conversation), or otherwise bombarded with irrelevant information, the driver may still be able to receive and process tactile warnings to aid perception of situation-relevant stimuli, thus improving SA and safety. In fact, Fitch et al. (2007) recently tested an automobile seat pan equipped with haptic (i.e., vibro-tactile) cueing. Used as a crash threat alerting mechanism to indicate the direction from which a crash may be imminent, the addition of the vibro-tactile cueing improved localization accuracy to 84% (compared to 34% for auditory cueing only) and reduced localization response times by 257 ms. The associated improvement in SA and reduced response times could mean the difference between a fatal crash and accident avoidance

Behavioral and Psychological Approaches

Whereas most approaches to improving SA have been task-specific and/or focused on increasing the saliency of environmental cues through warning systems, Mikulas (2007) suggested that mindfulness training could be more beneficial to attaining SA by improving one's overall awareness. Mikulas defined mindfulness as a behavior of the mind that involves "the active maximizing of the breadth and clarity of awareness" (p. 15). Langer (1997) suggested that mindfulness is composed of openness to novelty, alertness to distinction, sensitivity to different contexts, awareness of multiple perspectives and orientation in the present, all of which pertain to SA. Cultivating mindfulness can provide an individual with greater awareness of one's sensations, thoughts, and environmental cues. A related concept, concentration, can be thought of as the quieting of the mind and the controlled focusing of one's attention (Mikulas, 2007). Though mindfulness and concentration are two distinct constructs, they may be thought of as symbiotic behaviors of the mind that when both are cultivated help to improve one another. Both may be improved through meditation and other Buddhist practices. By being aware of when distracting thoughts are entering one's mind and having greater control in preventing them from doing so, mindful drivers should be less susceptible to inattention blindness. Rutschman (2004) found that in a divided-attention task, meditators and individuals practicing relaxation had greater success on the two simultaneous tasks (i.e., visual tracking and auditory stop tasks). Both groups showed improvement after meditation or relaxation practice, but meditation led to an increased flexibility in attention and greater sustainability in attentional processes than relaxation alone. A possible implication of this study is that meditation could enable drivers to more productively allocate resources of attention in proportion to the demands of driving and another task (e.g., conversation), thus increasing drivers' accident avoidance capability through improved perception and comprehension (levels 1 & 2 SA) of environmental stimuli

Currently scant literature is available on the connection between mindfulness training and SA, but such research could prove to be quite valuable for its potential benefits to driver safety. For example, it has been shown in a variety of situations that mindfulness results in a greater sensitivity to one's environment, which is particularly important when driving in monotonous or unfamiliar conditions. Mindfulness also has a strong connection to attention and attention improves when individuals are mindful of new stimuli in the environmental contexts (Langer & Moldoveanu, 2000). Increased attention in the driving environment can decrease the probability of risky driving and accidents.

Baer (2007) suggested that by practicing mindfulness one becomes better at directing attention to where it is needed the most. For example, she proposed that the potential benefit of mindfulness training for airline pilots is that they would be better able to direct their attention away from problematic thoughts and feelings (e.g., stress or fatigue), instead focusing on the aviation task at hand. Generalizing to drivers, whereas emotional or engaging thoughts may divert one's attention away from the road, by being aware of what the mind is trying to do, the mindful driver can avoid the distraction and redirect attention onto the relevant elements of driving, thereby improving SA.

Some studies have shown the effect mindfulness has on the performance of concentration tasks and attentional processes. These effects include decreased mental fatigue, and improved information processing and reaction time. For example, Barnes, Treiber and Davis (2001) evaluated the effects of transcendental meditation in a group of adolescents with high normal blood pressure. Utilizing a driving simulator as a stress inducer, at postintervention, stress responses were smaller for the group practicing meditation and consisted of decreases in systolic blood pressure, heart rate and cardiac output during driving sessions. Reasons for the stress reductions are unknown, but are assumed to be due to perceptual-motor coordination and/or faster reaction times associated with transcendental meditation. Improvements in perceptual-motor coordination and reaction time should greatly increase driver safety in terms of both controlling the vehicle and providing more time to make good driving decisions. Being able to control one's stress levels should have far-reaching effects on improving concentration, SA, and performance. In a related study, Matthews, Dorn, Hoyes, Davies, Glendon, and Taylor (1998) found that stress vulnerability adversely impacted driver performance. They suggested that stress management techniques would be particularly beneficial in helping drivers become more aware of their negative emotions and teach them how to deal with feelings of anger and frustration. Future research might investigate whether alternatives to technological approaches, such as mindfulness training, could improve SA and thus provide a better safety intervention for already technologically-overwhelmed drivers. Perhaps such training may be delivered as part of traditional driver education programs

CONCLUSION

This chapter examined the impact of the cellular phone in terms of drivers' loss of SA. The empirical evidence gathered over recent years leaves little doubt that cellular phones adversely impact driver performance, whether the phone is handheld or hands-free. However, lawmakers have largely ignored this evidence as only a half dozen states ban hand-held cellular phones, while the use of hands-free phones by drivers is allowed in all 50 states. Cellular phones, and other in-vehicle distractions, force drivers to divide their limited attentional resources among tasks, thereby focusing less of their attention on driving. With their attention diverted by conversation, drivers cannot perceive all pertinent environmental cues (level 1 SA), comprehend their meaning (level 2 SA), or make accurate projections about the future status of their vehicle and road conditions (level 3 SA). Though the authors advocate for tougher legal restrictions on the use of cellular phones (including hands-free) among drivers, we also recognize that the use of this technology, and other distractions, is on the rise. Fortunately, some researchers have been investigating other methods for improving SA and reducing the risk of accidents. Technological advances such as force-feedback and vibro-tactile warning systems have shown some early promise in promoting drivers' SA (or at least compensating for a lack of SA). Most of these methods are still in the experimental phase and typically work by alerting and directing drivers to potential dangers in the environment. The use of haptic displays is particularly intriguing because they provide information via a tactile medium, rather than through the drivers' often overloaded visual or auditory channels. However, behavioral and psychological approaches to improving SA may prove to be at least as effective as these technological advances. For example, mindfulness training may improve SA by teaching drivers how to be more aware of external and internal stimuli relevant to driving, stress, and distractions to attention. With increased awareness one can more effectively utilize self-control skills to increase attention, reduce distraction, and/or relax. Such an approach has the benefits of working from within the driver, not adding any cost to the vehicle or ever suffering mechanical failure, and providing the driver with useful coping strategies that could improve one's quality of life. Further investigation of these and other strategies for improving SA and safety are very much needed, particularly as the roads become increasingly filled with distracted drivers

REFERENCES

Associated Press (2007, July 15). Text messaging cited in crash that killed 5 teens

Baer, R. A. (2007). Mindfulness and preparation for flight. *Ergonomics in Design, 15(2)*, 21-22

Baddeley, A. D. (1986). *Working memory*. Oxford, UK: Oxford University Press

Baddeley, A. D. & Hitch, G. (1974). Working memory. In G. Bower (Ed.), *The psychology of learning and motivation, 8,* 47-90. New York: Academic Press

Barnes, V. A., Treiber, F. A. & Davis, H. (2001). Impact of transcendental meditation on cardiovascular function at rest and during acute stress in adolescents with high normal blood pressure. *Journal of Psychosomatic Research, 51,* 597-605.

Beede, K. E. & Kass, S. J. (2006). Engrossed in conversation: The impact of cell phones on simulated driving performance. *Accident Analysis and Prevention, 38,* 415-421

Cellular Telecommunications and Internet Association (2007). Wireless quick facts, December 2006. Retrieved August 13, 2007, from http://ctia.org/media/index.cfm/AID/10323.

Chase, W. G. & Simon, H. A. (1973). Perception in chess. *Cognitive Psychology, 4,* 55-81.

Consiglio, W., Driscoll, P., Witte, M. & Berg, W. P. (2003). Effect of cellular telephone conversations and other potential interference on reaction time in a braking response. *Accident Analysis and Prevention, 35,* 495-500

Diamond, D. D., Kass, S. J., Andrasik, F., Raj, A. K. & Rupert, A. H. (2002). Vibrotactile cueing as a master caution system for visual monitoring. *Human Factors and Aerospace Safety, 2,* 339-354.

Drews, F. A., Pasupathi, M. & Strayer, D. L. (2004). Passenger and cell-phone conversations in simulated driving. *Proceedings of the Human Factors and Ergonomics Society 48th Annual Meeting*, 2210-2212.

Durso, F. T. & Gronlund, S. D. (1999). Situation awareness. In F. T. Durso, R. S. Nickerson, R. W. Schvaneveldt, S. T. Dumais, D. S. Lindsay, & M. T. H. Chi (Eds.), *Handbook of applied cognition* (pp. 283-314). New York: Wiley.

Endsley, M. R. (1988). Design and evaluation for situation awareness enhancement. *Proceedings of the Human Factors Society 32nd Annual Meeting*, 97-101.

Endsley, M. R. (1995). Measurement of situation awareness in dynamic systems. *Human Factors, 37*, 65-84.

Ellis, R. D. (2001). Implications of inattentional blindness for "enactive" theories of consciousness. *Brain and Mind, 2,* 297-322.

Eysenck, M. W. & Keane, M. T. (1990). *Cognitive psychology: A student's handbook.* London: Lawrence Erlbaum Associates.

Fitch, G. M., Kiefer, R. J., Hankey, J. M. & Kleiner, B. M. (2007). Toward developing an approach for alerting drivers to the direction of a crash threat. *Human Factors, 49,* 710-790.

Foo, P., Warren, W. & Duchon, A. (2005). Do humans integrate routes into a cognitive map? Map- versus landmark-based navigation of novel shortcuts. *Journal of Experimental Psychology: Learning, Memory, and Cognition, 31,* 195-215.

Garcia-Larrea, L., Perchet, C., Perrin, F. & Amenedo, E. (2001). Interference of cellular phone conversations with visuomotor tasks: An ERP study. Journal *of Psychophysiology, 15,* 14-21.

Glassbrenner, D. (2005). *Driver cell phone use in 2004-Overall results* (DOT Publication No. DOT-HS-809-847). Washington, DC: National Center for Statistical Analysis.

Glassbrenner, D. & Ye, T. J. (2007). *Driver cell phone use in 2006-Overall results* (DOT Publication No. DOT-HS-810-790). Washington, DC: U.S. Department of Transportation, National Highway Safety Administration.

Gugerty, L. J. (1997). Situation awareness during driving: Explicit and implicit knowledge in dynamic spatial memory. *Journal of Experimental Psychology: Applied, 3,* 42-66.

Haigney, D. E., Taylor, R. G. & Westerman, S. J. (2000). Concurrent mobile (cellular) phone use and driving performance: Task demand characteristics and compensatory processes. *Transportation Research Part F: Traffic Psychology and Behaviour, 3,* 113-121.

Hopp, P. J., Smith, C. A. P., Clegg, B. A. & Heggestad, E. D. (2005). Interruption management: The use of attention-directing tactile cues. *Human Factors, 47,* 1-11.

Horrey, W. J. & Wickens, C. D. (2006). Examining the impact of cell phone conversations on driving using meta-analytic techniques. *Human Factors, 48,* 196-205.

Horswill, M. S., Waylen, A. E. & Tofield, M. I. (2004). Drivers' ratings of different components of their own driving skill: A greater illusion of superiority for skills that relate to accident involvement. *Journal of Applied Social Psychology, 34,* 177-194.

Insurance Institute for Highway Safety (2007). Cell phones and driving. Retrieved August 4,2007, from http://www.iii.org/;media/hottopics/insurance/cellphones/

Kass, S. J., Cole, K. S. & Stanny, C. J. (2007). Effects of distraction and experience on situation awareness and simulated driving. *Transportation Research Part F: Traffic Psychology and Behaviour, 10,* 321-329.

Kass, S. J., Herschler, D. A. & Companion, M. A. (1991). Training situation awareness through pattern recognition in a battlefield environment. *Military Psychology, 3,* 105-112.

Klauer, S. G., Dingus, T. A., Neale, V. L., sudweeks, J. D. & Ramsey, D. J. (2006). *The impact of driver inattention on near-crash/crash risk: An analysis using the 100-car naturalistic driving study data.* (Report No. DOT-HS-810-594). Washington, DC: U.S. Department of Transportation, National Highway Traffic Safety Administration.

Kozak, K., Pohl, J., Birk, W., Greenberg, J., Artz, B., Blommer, M., et al. (2006). Evaluation of lane departure warnings for drowsy drivers. *Proceedings of the Human Factors and Ergonomics Society 50th Annual Meeting,* 2400-2404.

Langer, P., Holzner, B., Magnet, W. & Kopp, M. (2005). Hands-free mobile phone conversation impairs the peripheral visual system to an extent comparable to an alcohol level of 4-5 g 100 ml. *Human Psychopharmacology: Clinical & Experimental, 20,* 65-66.

Langer, E. J. & Moldoveanu, M. (2000). The construct of mindfulness. *Journal of Social Issues, 56,* 1-9.

Ma, R. & Kaber, D. B. (2006). Situation awareness and driving performance in a simulated navigation task. *Proceedings of the Human Factors and Ergonomics Society 50th Annual Meeting,* 270-274.

Mack, A. (2003). Inattentional blindness: Looking without seeing. *Current Directions in Psychological Science, 12,* 180-184.

Mack, A., Pappas, Z., Silverman, M. & Gay, R. (2002). What we see: Inattention and the capture of attention by meaning. *Consciousness & Cognition: An International Journal, 11,* 488-506.

Manser, M. P., Ward, N. J., Kuge, N. & Boer, E. R. (2004). Influence of a driver support system on situation awareness and information processing in response to lead vehicle braking. *Proceedings of the Human Factors and Ergonomics Society 48th Annual Meeting,* 2359-2363.

Mathews, R., Legg, S. & Charleton, S. (2003). The effect of cell phone type on drivers' subjective workload during concurrent driving and conversing. Accident *Analysis and Prevention, 35,* 451-457.

Matthews, G., Dorn, L., Hoyes, T. W., Davies, D. R., Glendon, A. I. & Taylor, R. G. (1998). Driver stress and performance on a driving simulator. *Human Factors, 40,* 136-149.

May, J. F., Baldwin, C. L. & Parasuraman, R. (2006). Prevention of rear-end crashes in drivers with task-induced fatigue through the use of auditory collision avoidance warnings. *Proceedings of the Human Factors and Ergonomics Society 50th Annual Meeting,* 2409-2413

McKnight, A. J. & McKnight, A. S. (1993). The effect of cellular phone use upon driver attention. *Accident Analysis and Prevention, 25,* 259-265.

Mikulas, W. L. (2007). Buddhism & western psychology: Fundamentals of integration. *Journal of Consciousness Studies, 14 (4),* 4-49.

Patten, V. Kercher, A., Ostlund, J. & Nilsson, L. (2004). Using mobile telephones: Cognitive workload and attention resource allocation. *Accident Analysis and Prevention, 36,* 341-350.

Peters, G. A. & Peters, B. J. (2002). The distracted driver. *Journal of the Royal Society for the Promotion of Health, 121,* 23-28.

Posner, M. L. (1980). Orienting of attention. Quarterly *Journal of Experimental Psychology, 32,* 3-25.

Raj, A. K., Kass, S. J. & Perry, J. F. (2000). Vibrotactile displays for improving spatial awareness. *Proceedings of the International Ergonomics Association/Human Factors and Ergonomics Society Congress, 1,* 181-184.

Recarte, M. A. & Nunes, L. M. (2002). Mental load and loss of control over speed in real driving: Towards a theory of attentional speed control. *Transportation Research Part F: Traffic Psychology and Behaviour, 5,* 111-122.

Recarte, M. A. & Nunes, L. M. (2003). Mental workload while driving: Effects on visual search, discrimination, and decision making. *Journal of Experimental Psychology: Applied, 9,* 119-137.

Redelmeier, D. A. & Tibshirani, R. J. (1997). Association between cellular-telephone calls and motor vehicle accidents. *New England Journal of Medicine, 336,* 453-458.

Rutschman, J. R. (2004). Effects of techniques of receptive meditation and relaxation on attentional processing. *Canadian Undergraduate Journal of Cognitive Science,* 6-16.

Strayer, D. L. & Drews, F. A. (2007). Cell-phone-induced driver distraction. *Current Directions in Psychological Science, 16,* 128-131.

Strayer, D. L. & Johnston, W. (2001). Driven to distraction: Dual-task studies of simulated driving and conversing on a cellular phone. *Psychological Science, 12,* 462-466.

Strayer, D. L., Drews, F. A. & Crouch, D. J. (2006). A comparison of the cell phone driver and the drunk driver. *Human Factors, 48,* 381-391.

Strayer, D. L., Drews. F. A. & Johnston, W. A. (2007). Cell phone induced failures of visual attention during simulated driving. *Journal of Experimental Psychology: Applied, 9,* 23-32.

Treffner, P. J. & Barrett, R. (2004). Hands-free mobile phone speech while driving degrades coordination and control. *Transportation Research Part F: Traffic Psychology and Behaviour, 7,* 229-246.

Utter, D. (2001). *Passenger vehicle driver cell phone use: Results from the fall 2000 National Occupant Protection Use Survey* (DOT Publication No. DOT-HS-809-293). Washington, DC: U.S. Department of Transportation

Wickens, C. D. (1984). Processing resources in attention. In R. Parasuraman & R. Davies (Eds.), *Varieties of Attention.* (pp. 63-101). New York: Academic Press.

Wickens, C. D., Lee, J. D., Liu, Y. & Gordon-Becker, S. E. (2004). *An introduction to human factors engineering, 2nd Edition.* Upper Saddle River, NJ: Pearson Prentice Hall.

INDEX

A

B

C

D

E

P

Q

R

S

T

V

W

Y